# Algorithms for Intelligent Systems

This book series publishes research on the analysis and development of algorithms for intelligent systems with their applications to various real world problems. It covers research related to autonomous agents, multi-agent systems, behavioral modeling, reinforcement learning, game theory, mechanism design, machine learning, meta-heuristic search, optimization, planning and scheduling, artificial neural networks, evolutionary computation, swarm intelligence and other algorithms for intelligent systems.

The book series includes recent advancements, modification and applications of the artificial neural networks, evolutionary computation, swarm intelligence, artificial immune systems, fuzzy system, autonomous and multi agent systems, machine learning and other intelligent systems related areas. The material will be beneficial for the graduate students, post-graduate students as well as the researchers who want a broader view of advances in algorithms for intelligent systems. The contents will also be useful to the researchers from other fields who have no knowledge of the power of intelligent systems, e.g. the researchers in the field of bioinformatics, biochemists, mechanical and chemical engineers, economists, musicians and medical practitioners.

The series publishes monographs, edited volumes, advanced textbooks and selected proceedings.

**Indexed by zbMATH.**

**All books published in the series are submitted for consideration in Web of Science.**

Dhan Jeet Singh · Teena Sharma ·
Nishchal K. Verma

# Fuzzy Systems with Applications to Modeling and Control of an Aircraft

Springer

Dhan Jeet Singh
Aircraft Research and Design Centre
Hindustan Aeronautics Limited
Bengaluru, Karnataka, India

Teena Sharma
Mehta Family School of Data Science and AI
Indian Institute of Technology Guwahati
Guwahati, Assam, India

Nishchal K. Verma
Department of Electrical Engineering
Indian Institute of Technology Kanpur
Kanpur, Uttar Pradesh, India

ISSN 2524-7565 ISSN 2524-7573 (electronic)
Algorithms for Intelligent Systems
ISBN 978-981-95-7887-0 ISBN 978-981-95-7888-7 (eBook)
https://doi.org/10.1007/978-981-95-7888-7



This Springer imprint is published by the registered company Springer Nature Singapore Pte Ltd.
The registered company address is: 152 Beach Road, #21-01/04 Gateway East, Singapore 189721, Singapore

# Preface

Engineering systems operating in real-world environments are often characterized by uncertainty, imprecision, and nonlinear dynamics. Traditional mathematical models, while powerful, frequently struggle to fully capture the complexity of such systems when uncertainty arises from incomplete information, noisy measurements, or linguistic knowledge provided by human experts. Fuzzy logic, introduced by Lotfi A. Zadeh in 1965, offers a powerful framework for modeling and reasoning under such uncertain conditions. Over the past several decades, fuzzy systems have evolved significantly, progressing from Type-1 fuzzy systems to more sophisticated structures such as Interval Type-2, General Type-2, and, more recently, Type-3 fuzzy systems, each providing enhanced capabilities for handling uncertainty.

This book presents a comprehensive study of fuzzy systems with a particular emphasis on their application to the modeling and control of aircraft dynamics. Aircraft systems represent highly nonlinear and complex dynamical systems that operate under a wide range of environmental conditions and operational constraints. These systems often experience uncertainties arising from aerodynamic parameter variations, modeling inaccuracies, actuator limitations, and external disturbances. Fuzzy systems provide an effective methodology for addressing these challenges by integrating both objective mathematical models and subjective expert knowledge within a unified framework.

The primary motivation of this book is to explore advanced fuzzy system methodologies and demonstrate their effectiveness in aerospace control applications. In particular, the book investigates the design and analysis of Type-1, Interval Type-2, General Type-2, and Interval Type-3 fuzzy systems, with a focus on their robustness, modeling capability, and applicability in flight control systems. Through theoretical development and practical demonstrations, the book illustrates how higher-order fuzzy systems can enhance system performance when dealing with uncertain and nonlinear environments.

The book begins with an introduction to fuzzy logic theory and its evolution, highlighting the need for higher-order fuzzy systems capable of representing multiple layers of uncertainty. It then provides a detailed discussion on the development of fuzzy sets and inference systems, followed by the design of fuzzy model-based control systems for aircraft applications. Particular attention is given to Takagi-Sugeno (T-S) fuzzy model-based controllers, which enable the approximation of nonlinear aircraft dynamics through multiple local linear models combined using fuzzy rules.

Subsequent chapters extend the discussion toward Interval Type-2 fuzzy control systems, which introduce footprints of uncertainty in membership functions, allowing more effective handling of modeling errors and environmental variations. The book also presents a systematic methodology for analyzing the robustness of General Type-2 fuzzy systems, highlighting their advantages in managing noisy and uncertain data. Furthermore, the book introduces and explores the emerging concept of Interval Type-3 fuzzy systems, which provide an additional level of uncertainty representation and improved modeling flexibility.

A key contribution of this book is the application of these advanced fuzzy methodologies to aerospace modeling and flight control problems. This book is intended for graduate students, researchers, and practicing engineers working in the fields of artificial intelligence, control systems, aerospace engineering, and computational intelligence. It may also serve as a useful reference for those interested in uncertainty modeling and advanced fuzzy system design. The authors aim to provide both theoretical insights and practical guidance for the development of robust fuzzy-based solutions in complex dynamical systems.

The authors would like to express their sincere gratitude to colleagues, collaborators, and students who have contributed to the research work presented in this book. Their discussions, suggestions, and support have played a significant role in shaping the ideas and methodologies described here. We also acknowledge the broader research community whose foundational contributions in fuzzy logic and intelligent control systems have inspired this work.

The authors hope that this book will contribute to advancing research in fuzzy systems and encourage further exploration of higher-order fuzzy logic in engineering applications.

Bengaluru, India
Guwahati, India
Kanpur, India
March 2026

Dhan Jeet Singh
Teena Sharma
Nishchal K. Verma

# Contents

# About the Authors

**Dr. Dhan Jeet Singh** received his Ph.D. degree in Electrical Engineering from the Indian Institute of Technology Kanpur, India. Dr. Singh is currently working with Aircraft Research and Design Centre, Hindustan Aeronautics Limited (HAL), Bengaluru, India. Dr. Singh's research interests are Artificial Intelligence, Machine Learning, Deep Learning, and Higher-Order Fuzzy Systems and Control. Dr. Singh is member of Aeronautical Society of India (AeSI), Institute of Engineers (IEI) India and Institute of Electrical and Electronics Engineers (IEEE). Dr. Singh has been serving as active reviewer of various reputed journals and conferences which include IEEE Transactions on Artificial Intelligence, IEEE Transactions on Neural Networks and Learning Systems, and many more.

**Dr. Teena Sharma** is an Assistant Professor at the Mehta Family School of Data Science and Artificial Intelligence (MFSDS&AI), Indian Institute of Technology Guwahati, India. Dr. Sharma worked as a Postdoctoral Scholar at the University of Tennessee, Memphis, Tennessee, USA. Dr. Sharma received her Ph.D. degree in Electrical Engineering from the Indian Institute of Technology Kanpur, India. Dr. Sharma's research interests are Artificial Intelligence, Machine Learning, and Deep Learning Algorithms and their applications to Computer Vision: Object detection, Classification, Identification, Recognition, Image enhancement, Image matching; Equitable Precision Medicine: Transfer learning, Meta-learning, Few-shot learning; and Condition-based Monitoring: Fault diagnosis and remaining useful life prediction. Dr. Sharma is also serving as an Associate Editor for the IEEE Transactions on Artificial Intelligence.

**Dr. Nishchal K. Verma** is a Professor (HAG) of Electrical Engineering at the Indian Institute of Technology Kanpur, India, and a leading researcher in artificial intelligence and computational intelligence. His research focuses on AI theory and its realworld deployment across interdisciplinary domains, including machine learning, deep and hybrid models, computer vision, prognostics and health management (PHM), cyberphysical systems, bioinformatics, and complex nonlinear system modeling.

Dr. Verma has published 275+ peerreviewed papers and four books and has led 23 funded research projects supported by premier agencies and industries such as The Boeing Company (USA), DST, DRDO, SERB, CSIR, MHRD, and IIT Kanpur. His work has contributed to both fundamental advances and largescale industrial applications of AI. He serves on the editorial boards of leading journals.

# List of Figures

# List of Tables

# Chapter 1
# Introduction

**Abstract** Fuzzy logic serves as a crucial tool for managing uncertainty, imprecision, and vagueness, offering a flexible mathematical framework for modeling and reasoning in situations where traditional binary logic (true/false) falls short in capturing real-world complexity. Fuzzy logic systems are classified according to their capacity to handle uncertainty within their membership functions. The progression from Type-1 fuzzy logic system to Type-2 and Type-3 reflects increasing sophistication in uncertainty modeling, allowing for a more accurate representation of ambiguous information. Type-1 fuzzy logic system operates with precise membership functions, which assume no uncertainty in the definition of fuzzy sets. However, real-world environments often involve higher levels of imprecision, where this assumption breaks down. Interval Type-2 fuzzy logic system addresses this limitation by incorporating uncertainty directly into its membership functions, allowing for a range of possible values rather than fixed ones. General Type-2 fuzzy logic system extends this even further, providing a more comprehensive way to model and control systems in highly uncertain environments. This chapter introduces the motivation behind and potential applications of General Type-2 and Interval Type-2 fuzzy logic systems for the modeling and control of aircraft, where high precision and adaptability to changing conditions are critical.

## 1.1 Introduction

In many engineering applications, there exist two different forms of problem knowledge. The first one is "objective knowledge", and the second one is "subjective knowledge". Objective knowledge is signified by mathematical models and equations, whereas subjective knowledge represents linguistic information, e.g., rules, expert information, design requirements, etc., that is difficult to quantify using traditional mathematics. The fuzzy logic-based theory can combine both forms of knowledge in a logical and coordinated way and can be utilized in various applications [1–3].

The growth of fuzzy sets advanced by adding a new dimension in the formulation of uncertainty. The Interval Type-2 (IT2) Fuzzy Logic System (FLS) has evolved from the basis of Type-1 (T1) FLS, which confines for uncertainties in the Membership

 
D. J. Singh et al., *Fuzzy Systems with Applications to Modeling and Control of an Aircraft*, Algorithms for Intelligent Systems, https://doi.org/10.1007/978-981-95-7888-7_1

Function (MF) and weights it uniformly [4–8]. Further, General Type-2 (GT2) FLS emerged as an extension of IT2 FLS, which also confines to uncertainties in MF, but these uncertainties are non-uniformly weighted [9, 10]. GT2 FLSs are well-suited for designing robust systems capable of managing vagueness and uncertainty effectively. To further enhance performance when dealing with uncertain data or information originating from sources with inherently uncertain states—where uncertainty persists even within uncertain conditions—research efforts have been extended to develop Interval Type-3 (IT3) Fuzzy Logic Systems as an advanced iteration of GT2 FLS.

## 1.2 Background

In many applications, the data or information used to analyze or model a system's characteristics often contains significant levels of vagueness and uncertainty. These uncertainties can be either random or linguistic in nature, with linguistic uncertainties typically associated with imprecise descriptions or words. Fuzzy systems theory provides a robust framework for addressing both types of uncertainties, enabling efficient operation even in the presence of random disturbances and noisy measurements [11].

Type-1 fuzzy model comprises fuzzy rule sets as a part of the fuzzy inference engine. In a few practical applications, the available information or data is so vague and uncertain that it becomes very difficult to construct the appropriate rule base. Thus, the rule base itself appears uncertain in nature. In such a situation, to achieve the desired level of performance or accuracy, the need for Type-2 FLS arises.

In [12], Mendel proposed that linguistic and random uncertainties can be encapsulated within the Footprint of Uncertainty (FOU) in IT2 FLS. Consequently, the "uncertainty" measurement pertains to quantifying the FOU. This feature of IT2 FLS distinguishes it from T1 FLS in effectively handling uncertainty. In IT2 FLS, the FOU is associated with the primary MF, while the secondary MF is constant with a value of one. Similarly, in GT2 FLS, the FOU is also tied to the primary MF, but the secondary MF itself has the structure of a T1 FLS. Building upon this theoretical foundation, IT3 FLS has been introduced [13], wherein the secondary MF follows the structure of an IT2 FLS. This results in a FOU linked to both the primary and secondary MFs, demonstrating the capability of IT3 FLS to model higher degrees of uncertainty through FOUs at multiple levels of the membership hierarchy.

In [14], Rickard et al. proposed that the MFs of Type-$n$ fuzzy sets take values within Type-($n$-1) fuzzy sets defined on the unit interval, enabling these sets to represent distinct origins of uncertainty. The concept of "uncertain value for an already uncertain value" has been illustrated with an example. In this case, the uncertainty related to the whiteness grade of a particular sample arises from factors such as variations in lighting conditions and differences in the fabric on which the color is printed. Consequently, the higher-type fuzzy sets, such as Type-3 or beyond, may be necessary to effectively model the concept of "white" in this context. Similarly, several cases highlighting the potential applicability of Type-3 FLS have been discussed in [15, 16].

On the application front, this book focuses on the aerospace modeling and control-related application of higher-order fuzzy systems. The motive behind these applications is the usefulness of fuzzy systems to embed subjective insight and statistical understanding of aircraft dynamics for various design applications. These design applications are centered on examining and evaluating the intricate nonlinear dynamics of aircraft, encompassing diverse stages, such as conceptualization and prototyping, servicing and upkeep, system modernization, performance enhancement, error monitoring, and formulation of control algorithms, along with other related aspects. Apart from this, uncertainties arose due to variations in aerodynamic parameters with changes in flight operating scenarios, inaccurate control execution, and unrecognized or undefined nonlinear behaviors of aircraft dynamics, which can be easily handled by fuzzy logic theory.

Managing uncertainty will be a critical aspect for the majority of Artificial Intelligence (AI)-based systems designed to address real-world challenges. AI community is using various theories as core research in the field of uncertainty management. These theories are fuzzy logic, belief function, and Bayesian probability theory [17–19]. In the design of FLS, there are two different methods, namely Type-1 (T1) and Type-2 (T2) FLS. The T2 FLS was projected by Prof. Zadeh to handle the high level of vagueness and uncertainty from the theory of T1 FLS as its advance expansion [20, 21]. From the literature, it is proved that higher-order fuzzy systems can offer more robustness to noisy data and handle unexpected patterns of signals, such as extreme values and outliers. Further, to attain more adequate performance while handling uncertain data or information that originated from uncertain sources, research on IT3 FLS is explored [13].

## 1.3 Objectives

The prime objectives of this book are to introduce methods for the design of Interval Type-3 FLS. The robustness analysis for GT2 FLS is also analyzed and introduced in this book. In addition to this, this book also illustrates a systematic technique for the design of an IT2 Takagi-Sugeno (T-S) Fuzzy Model Based (FMB) flight control system for Generic Aircraft/UAV-related applications. For this application, this book explores the concept of the mapping of flight envelope parameters as premise parameters of IT2 T-S FLS.

System modeling and controller design is a well-known application of fuzzy systems theory. Fuzzy controllers can provide robust performance, model-based design, nonlinear modeling ability, and other advanced features than classical alternatives. In addition to this, fuzzy controllers are very suitable for dealing with complex dynamical systems. It can handle the inherent uncertainty which arose due to the noise, dynamic changes, and imprecision in the models.

Modern aircraft is a well-known example of a highly nonlinear complex dynamical system in which there exist various kinds of uncertainty and unsteadiness due to disrupted airflow, structural/aerodynamic/propulsion couplings, variations in mass properties, and similar factors during flight. The IT3 T-S FLS can be applied for modeling of nonlinear quasi steady stall longitudinal aerodynamics by using the recorded flight data of Advanced Technologies Testing Aircraft System (ATTAS) aircraft.

## 1.4 Organization

Following the aforementioned introduction, background, and objectives, the contents of this book further introduce and explain the following.

This book presents an approach to the design of an Interval Type-3 (IT3) FLS. An IT3 FLS is designed with the baseline of the GT2 FLS, where the Secondary Membership Function (SMF) is an IT2 Fuzzy Set (FS), and values of tertiary membership are unity over the FOU of SMF. The introduced system is particularly advantageous for handling uncertain information or data originating from inherently uncertain sources, where layers of uncertainty persist, even within already uncertain conditions.

In the literature, the robustness analysis for IT2 FLS and Type-1 FLS already exists. This book introduces and explains the method for robustness analysis of GT2 FLSs. In the presented work for this purpose, the theoretical progression for GT2 FLSs is extended. Two simulation studies are presented to show the efficacy of the projected method. The result indicates that GT2 FLS can provide robust performance depending upon the type of application.

This book explores the application of an IT2 T-S Fuzzy Model Based (FMB) control system in the context of generic aircraft design. The IT2 T-S fuzzy logic system is derived by approximating the nonlinear dynamics of an aircraft around multiple representative equilibrium points within its flight envelope, governed by a set of fuzzy rules. In this model, the aircraft's flight envelope parameters, such as operational altitude and speed, serve as premise variables, while the stability and control derivatives matrix components function as consequent parameters. The longitudinal dynamics of the X-29A research aircraft are chosen as the basis for implementing the IT2 T-S FMB control system.

In continuation to this, another application of FMB control is also presented in this book for Generic Aircraft/UAVs. In this application, constraints on state and control input parameters are considered for designing the FMB control system. The Linear Matrix Inequalities (LMI) stability conditions are evaluated to generate the state feedback control gains.

## 1.5 Summary

This chapter provides an overview of FLS and their evolution from Type-1 to more advanced types, such as IT2 and GT2 FLS, with a future outlook on IT3 FLS. It highlights the significance of fuzzy logic in addressing uncertainty, vagueness, and imprecision, which are prevalent in many real-world scenarios, particularly in dynamic and complex environments like aerospace systems. The chapter emphasizes how IT2 and GT2 FLSs extend the basic FLS framework to handle higher levels of uncertainty and improve the robustness and precision required for applications such as aircraft modeling and control. The use of fuzzy logic in integrating both objective (mathematical) and subjective (linguistic) knowledge is underscored, positioning fuzzy logic as a crucial tool in engineering and AI-based systems that demand adaptability and robustness.

## References


1. T. Sharma and N. K. Verma, "Adaptive interval Type-2 fuzzy filter: an AI agent for handling uncertainties to preserve image naturalness," *IEEE Transactions on Artificial Intelligence*, vol. 2, no. 1, pp. 83–92, 2021. https://doi.org/10.1109/TAI.2021.3077522.
2. T. Sharma and N. K. Verma, "Estimating depth and global atmospheric light for image dehazing using Type-2 fuzzy approach," *IEEE Transactions on Emerging Topics in Computational Intelligence*, vol. 6, no. 1, pp. 93–102, 2022. https://doi.org/10.1109/TETCI.2020.3032970.
3. T. Sharma, N. K. Verma, and S. Masood, "Mixed fuzzy pooling in convolutional neural networks for image classification," *Multimedia Tools and Applications*, vol. 82, no. 6, pp. 8405–8421, 2023. https://doi.org/10.1007/s11042-022-13553-0.
4. O. Castillo and P. Melin, *Type-2 Fuzzy Logic: Theory and Applications*. Heidelberg: Springer, 2008.
5. J. M. Mendel, H. Hagras, W. W. Melek, and R. I. John, "Design of Interval Type-2 Fuzzy Logic Systems," *IEEE Transactions on Fuzzy Systems*, 2014.
6. J. M. Mendel, "Uncertainty and Fuzziness in Type-2 Fuzzy Logic Systems," *Information Sciences*, 2017.
7. S. M. Mirjalili, O. Castillo, and P. Melin, "A Review of Interval Type-2 Fuzzy Logic Systems and Their Applications," *Applied Soft Computing*, 2021.
8. J. M. Mendel and D. Wu, *Interval Type-2 Fuzzy Sets and Systems: Theory and Applications*. Heidelberg: Springer, 2010.
9. J. M. Mendel and R. I. Bobadilla, "An Introduction to General Type-2 Fuzzy Logic Systems," *IEEE Computational Intelligence Magazine*, 2020.
10. H. Wu, L. Dai, and J. M. Mendel, "General Type-2 Fuzzy Systems for Control of Nonlinear Systems," *IEEE Transactions on Fuzzy Systems*, 2013.
11. J. M. Mendel, *Uncertain Rule-Based Fuzzy Logic Systems: Introduction and New Directions*. Upper Saddle River, NJ: Prentice-Hall, 2001.
12. J. M. Mendel, "A quantitative comparison of interval Type-2 and Type-1 fuzzy logic systems: First results," in *Proceedings of the International Conference on Fuzzy Systems*, 2010, pp. 1–8. https://doi.org/10.1109/FUZZY.2010.5584727.
13. D. J. Singh and N. K. Verma, "Interval Type-3 T-S fuzzy system for nonlinear aerodynamic modeling," *Applied Soft Computing*, vol. 150, p. 111097, 2024. https://doi.org/10.1016/j.asoc.2023.111097.
14. J. T. Rickard, J. Aisbett, G. Gibbon, and D. Morgenthaler, "Fuzzy subsethood for Type-n fuzzy sets," in *NAFIPS 2008 - 2008 Annual Meeting of the North American Fuzzy Information Processing Society*, 2008, pp. 1–6. https://doi.org/10.1109/NAFIPS.2008.4531276.
15. J. T. Rickard, J. Aisbett, and G. Gibbon, "Fuzzy subsethood for fuzzy sets of Type-2 and generalized Type-n," *IEEE Transactions on Fuzzy Systems*, vol. 17, no. 1, pp. 50–60, 2009. https://doi.org/10.1109/TFUZZ.2008.2006369.

16. P. Fisher, T. Cheng, and J. Wood, "Higher order vagueness in geographical information: Empirical geographical population of Type-n fuzzy sets," *GeoInformatica*, vol. 11, no. 3, pp. 311–330, 2007. https://doi.org/10.1007/s10707-006-0009-5.
17. L. A. Zadeh, "Fuzzy sets," *Information and Control*, vol. 8, no. 3, pp. 338–353, 1965.
18. A. P. Dempster and G. Shafer, *A Mathematical Theory of Evidence*. Princeton, NJ: Princeton University Press, 1976.
19. J. Pearl, *Probabilistic Reasoning in Intelligent Systems: Networks of Plausible Inference*. Morgan Kaufmann, 1988.
20. L. A. Zadeh, "The concept of a linguistic variable and its application to approximate reasoning," *Information Sciences*, vol. 8, no. 3, pp. 199–249, 1975.
21. J. M. Mendel, *Uncertain Rule-Based Fuzzy Logic Systems: Introduction and New Directions*. Upper Saddle River, NJ: Prentice-Hall, 2001.

# Chapter 2
# Advancements in Fuzzy Logic

**Abstract** This chapter presents a thorough review of the latest advancements in fuzzy systems, with a particular emphasis on the progression from Type-1 fuzzy sets to Interval Type-2 and General Type-2 fuzzy sets. Starting with the foundational work by Prof. Lofti A. Zadeh in 1965, the chapter discusses the key advancements in fuzzy logic theory, including the introduction of fuzzy algorithms, decision-making processes, and their extensive applications in control engineering, pattern recognition, and artificial intelligence. Special attention is given to the theoretical aspects of Type-1 fuzzy sets, which serve as the foundation for more advanced fuzzy models. The chapter explores the need for Interval and General Type-2 fuzzy sets to handle higher degrees of uncertainty and imprecision, discussing their structure, mathematical representations, and applications.

## 2.1 Introduction

In many design applications, there may be a high chance that available information or data is vague and imprecise. Sometimes, this information is available in fragments, along with ambiguity in the design data. In this situation, it becomes very difficult to analyze the system behavior or model it as the knowledge available for this purpose becomes uncertain.

The uncertainties may be random (unpredictability) or linguistic. In [1], Klir and Wierman categorized uncertainty into

- Vagueness: lack of sharp boundaries in concepts or definitions.
- Imprecision: incomplete or insufficiently precise information.
- Strife/ Disorder: conflicting or contradictory information.

Aligning with Klir and Wierman, Berenji [2] stated that *Uncertainty can originate from various factors, including incompleteness, imprecision, randomness, or missing information.* Additionally, fuzzy systems are sometimes calibrated using unreliable or erroneous data, leading to yet another form of uncertainty that can further complicate system analysis and design.

 
D. J. Singh et al., *Fuzzy Systems with Applications to Modeling and Control of an Aircraft*, Algorithms for Intelligent Systems, https://doi.org/10.1007/978-981-95-7888-7_2

The fuzzy logic methodology offers solutions to tackle uncertainties (random or linguistic) and can function efficiently under disturbances [3]. In the design of FLS, there are two distinct techniques, namely T1 FLS and T2 FLS. The T2 FLS was projected by Prof. Zadeh to handle the high level of vagueness and uncertainty from the conception of T1 FLS as its advanced development. The MF of T1 FLS is crisp or certain, whereas the MF of T2 FLS is expressed as Secondary MF (SMF). In IT2 FLS, the SMF remains a constant value, but in GT2 FLS, it is derived from the T1 fuzzy set, thus providing more flexibility [4, 5]. After the initiation of the theory of fuzzy systems in 1965, it went through various significant developments in theoretical as well as application-oriented perspectives. There exists a large variety of fuzzy theory applications in control, pattern recognition, signal processing, robotics, finance, etc.

## 2.2 Fuzzy Systems: Historical Development

The theory of fuzzy sets originated in 1965, proposed by Prof. Lotfi A. Zadeh [6]. The further concept of fuzzy algorithms was projected in 1970 by Bellman and Zadeh [7]. The seminal paper, [8], was published by them in 1973. This paper established the foundation of linguistic variables and conceptualized the formulation of IF-THEN fuzzy rules for human knowledge representation. The basic background of the fuzzy controller was founded in 1975 by Mamdani and Assilian, by controlling a steam engine. This work was published in a fuzzy theory seminal article [9]. The full-scale fuzzy control was developed in 1978 by Holmblad and Ostergaard for industrial processes [10]. The fuzzy control application in water purification plants was created by Sugeno in the 1980s [11]. Further, in 1985, Sugeno and Nishida applied fuzzy theory in command control of a self-parking car [12]. In 1989, Hirota, Arai, and Hachisu designed a two-dimensional fuzzy robot arm that could play in real time [13].

With the growing acceptance of fuzzy set theory, it was recognized by various engineering organizations. Further, considering the applicability and advantages associated with fuzzy systems theory, various researchers have contributed sufficiently in this field.

In the literature, the most popularly used fuzzy inference approaches are Takagi-Sugeno (T-S) and Mamdani Fuzzy Inference Systems (FIS). The major difference among these approaches is in the consequent portion of the fuzzy rules. A Mamdani FIS uses fuzzy sets as a rule consequent, while T-S FIS uses the linear function of input parameters as a rule consequent. Due to differences in the consequent portion of fuzzy rules, the aggregation and defuzzification procedures of both methods are different from each other [3, 14].

The FIS formulates the appropriate number of fuzzy rules, and based on this, the proper decision is generated. FIS is based on the notion of fuzzy IF-THEN statements, which use "OR" or "AND" connectors in the rules to make the necessary decisions. FIS comprises a fuzzification part, a rule base, a database, a decision-making unit, and a defuzzification part.

The compositional blocks of FIS, as shown in Fig. 2.1 are:

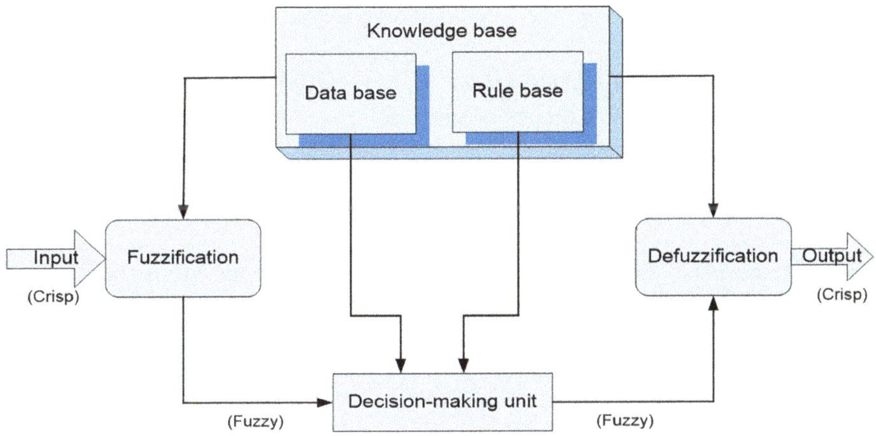


**Fig. 2.1** Fuzzy inference system

1. The database comprises MFs and their parameters.
2. The rule base comprising IF-THEN fuzzy rules.
3. The decision-making unit for rules inference processes.
4. The fuzzification interface for mapping the crisp inputs into the fuzzy domain with respective membership functions.
5. The defuzzification interface.

In FIS, the crisp input is first converted to the fuzzy domain through the fuzzification process, followed by the generation of the rule base. The rule base and database are collectively described as a knowledge base. Finally, the defuzzification step transforms the fuzzy value to a crisp output.

The stages of fuzzy inference processes as executed by FIS are:

- Compute the membership degree of each input variable mapped through appropriate membership functions on the premise part (Fuzzification step).
- Combine membership values of the premise part and obtain the firing strength (weight) corresponding to each rule through a t-norm (min) or co-norm (product) operator.
- Create the qualified consequents based on firing strength.
- Combined the qualified consequents to obtain a crisp output (defuzzification).

## 2.3 Type-1 Fuzzy Set

T1 FS denoted as $F$, is signified by T1 MF $\mu_F(x)$ where $x \in X$ with $X$ as the universe of discourse.

The set-theoretic representation of T1 FS is formulated as:

$$F = \{(x, \mu_F(x)) \mid \forall x \in X\} \tag{2.1}$$

where $\mu_F(x)$ denotes the grade of MF $(x, \mu_F(x)) \mid \forall x \in X$ with $0 \leq \mu_F(x) \leq 1$.

For continuous $X$, the T1 FS can be given as

$$F = \int_{x \in X} \mu_F(x)/x \tag{2.2}$$

where $\int$ represents union over all points $x \in X$ with associated MF $\mu_F(x)$.

For discrete $X$, the T1 FS $F$ can be given as

$$F = \sum_{x \in X} \mu_F(x)/x \tag{2.3}$$

In T1 FS, the intersection, union, and complements are denoted by using their MF. Let FSs $F_1$ and $F_2$ be defined by their MFs $\mu_{F_1}(x)$ and $\mu_{F_2}(x)$, then the MFs after the union and the intersection of these FSs are given by:

$$\mu_{F_1 \cup F_2}(x) = \max\left[\mu_{F_1}(x), \mu_{F_2}(x)\right] \tag{2.4}$$

and

$$\mu_{F_1 \cap F_2}(x) = \min\left[\mu_{F_1}(x), \mu_{F_2}(x)\right]. \tag{2.5}$$

Additionally, the fuzzy complement MF of $F$ is specified by

$$\mu_F(x) = 1 - \mu_F(x) \tag{2.6}$$

The subsequent section briefs about IT2 fuzzy sets as a further development in the T1 fuzzy set.

## 2.4 Interval Type-2 Fuzzy Sets

In general, T1 FSs are utilized for modeling across various domains; however, when uncertainty exists in the measurements or data, alternative approaches may be required like T2 FSs [3–5, 15].

As the membership function of T1 FSs is crisp in nature, it is unable to handle these uncertainties, whereas T2 FSs can model these uncertainties due to the fuzzy nature of their MFs. The MF of T2 FSs is three-dimensional, while the MF of T1 FSs is two-dimensional [5]. Hence, T2 FSs offer an extra level of flexibility to effectively represent and manage uncertainties. T2 FSs are advantageous for time-varying systems with unknown dynamics or non-stationary noise lacking a mathematical model [3].

Due to the computational complexity present in GT2 FSs, its application is limited. The IT2 FSs give the best compromise between the modeling accuracy of T2 FSs and the computational simplicity of T1 FSs. IT2 FSs are the simplified form of GT2 FSs with the secondary membership value of unity. This feature reduces the computational complexity associated with IT2 FSs.

In [4], Mendel et al. have presented the theoretical overview for IT2 FSs using various definitions and T1 FSs mathematics.

The T2 FSs, signified as $\tilde{F}$, is characterized by T2 MF $\mu_{\tilde{F}}(x, u)$, where $x \in X$, $u \in J_x \subseteq [0, 1]$, and $0 \leq \mu_{\tilde{F}}(x, u) \leq 1$, i.e.,

$$\tilde{F} = \{(x, u), \mu_{\tilde{F}}(x, u) | \forall x \in X, \forall u \in J_x \subseteq [0, 1]\}. \tag{2.7}$$

Another representation of $\tilde{F}$ is

$$\tilde{F} = \int_{x \in X} \int_{u \in J_x} \mu_{\tilde{F}}(x, u)/(x, u) \quad J_x \subseteq [0, 1] \tag{2.8}$$

The symbology $\int \int$ represents union over all permissible $x$ and $u$. In the case of the discrete $X$, the representation $\int$ is replaced by $\sum$.

When all $\mu_{\tilde{F}}(x, u) = 1$, then $\tilde{F}$ is an IT2 FS. Hence, IT2 FS is a specific instance of GT2 FS and Eq. (2.8) is represented as,

$$\tilde{F} = \int_{x \in X} \int_{u \in J_x} 1/(x, u) \quad J_x \subseteq [0, 1] \tag{2.9}$$

In [5], the concept of the vertical slice is presented with respect to the 2-D plane at $x = x'$. The 2-D plane with axes $u$ and $\mu_{\tilde{F}}(x', u)$ is termed as the vertical slice of $\mu_{\tilde{F}}(x, u)$. The secondary MF $\mu_{\tilde{F}}(x = x', u)$ for $x' \in X$ and $\forall u \in J_{x'} \subseteq [0, 1]$ is a vertical slice of $\mu_{\tilde{F}}(x, u)$ and mathematically expressed as

$$\mu_{\tilde{F}}(x') \equiv \mu_{\tilde{F}}(x = x', u) = \int_{u \in J_{x'}} 1/u \quad J_{x'} \subseteq [0, 1] \tag{2.10}$$

The uncertainty in the primary membership of $\tilde{F}$ is a bounded region and is known as the FOU. The FOU is the union of all primary memberships and is expressed as

$$FOU(\tilde{F}) = \bigcup_{x \in X} J_x \tag{2.11}$$

The upper and lower MFs (i.e., UMF and LMF) are T1 MFs that bound the FOU. The upper bound of $FOU(\tilde{F})$ is related with UMF and signified by $\overline{\mu}_{\tilde{F}}(x)$. Similarly, the lower bound of $FOU(\tilde{F})$ is related with LMF and signified by $\underline{\mu}_{\tilde{F}}(x)$, i.e.,

$$\overline{\mu}_{\tilde{F}}(x) \equiv FOU(\tilde{F}) \quad \forall x \in X \tag{2.12}$$

and

$$\underline{\mu}_{\tilde{F}}(x) \equiv FOU(\tilde{F}) \quad \forall x \in X. \tag{2.13}$$

For discrete $X$ and $U$, the embedded IT2 FS, $\tilde{F}_e$, has $N$ elements, namely $u_1, u_2, \ldots, u_N$ having unity secondary grade, i.e.,

$$\tilde{F}_e = \sum_{i=1}^{N} [1/u_i]/x_i, \quad u_i \in J_{x_i} \subseteq U = [0, 1] \tag{2.14}$$

where set $\tilde{F}_e$ is defined as the embedded set of $\tilde{F}$.

The wavy slice depiction of an IT2 FS $\tilde{F}$ is the aggregation of all its constituent embedded IT2 FSs, i.e.,

$$\tilde{F} = \sum_{j=1}^{n_r} \tilde{F}_e^j \tag{2.15}$$

$$\tilde{F}_e^j = \sum_{i=1}^{N} \left[1/u_i^j\right]/x_i, \quad u_i^j \in J_{x_i} \subseteq U = [0, 1] \tag{2.16}$$

Various set-theoretic operations concerning the "complement", "intersection", and "union" of IT2 FS are formulated as shown further. The MF value resulting from union of two IT2 FSs ($\tilde{F}_1$ and $\tilde{F}_2$) is,

$$\mu_{\tilde{F}_1 \cup \tilde{F}_2}(x) = \frac{1}{2}\left[\underline{\mu}_{\tilde{F}_1}(x) \vee \underline{\mu}_{\tilde{F}_2}(x), \overline{\mu}_{\tilde{F}_1}(x) \vee \overline{\mu}_{\tilde{F}_2}(x)\right] \quad \forall x \in X \tag{2.17}$$

The MF value resulting from intersection of two IT2 FSs ($\tilde{F}_1$ and $\tilde{F}_2$) is,

$$\mu_{\tilde{F}_1 \cap \tilde{F}_2}(x) = \frac{1}{2}\left[\underline{\mu}_{\tilde{F}_1}(x) \wedge \underline{\mu}_{\tilde{F}_2}(x), \overline{\mu}_{\tilde{F}_1}(x) \wedge \overline{\mu}_{\tilde{F}_2}(x)\right] \quad \forall x \in X \tag{2.18}$$

The MF value resulting after complement of IT2 FS is,

$$\mu_{\overline{\tilde{F}}} = \frac{1}{2}\left[(1 - \overline{\mu}_{\tilde{F}}(x)), (1 - \underline{\mu}_{\tilde{F}}(x))\right] \quad \forall x \in X \tag{2.19}$$

The next section briefs about GT2 FSs as a further development in IT2 FSs.

## 2.5 General Type-2 Fuzzy Sets

This section provides a brief overview of the theoretical characteristics of GT2 FSs. For completeness, the mathematical notations presented in [5, 17] are also reproduced.

The GT2 FS possesses a 3-D MF, where each primary MF is associated with a corresponding secondary MF. The general representation is as,

$$\tilde{F} = \{(x, u), \mu_{\tilde{F}}(x, u) | \forall x \in X, \forall u \in [0, 1]\} \tag{2.20}$$

where $x$ is the primary variable and $u$ is a secondary variable.

The secondary MF depends on both the primary and secondary variables and is represented using $\mu_{\tilde{F}}(x, u)$ with $0 \leq \mu_{\tilde{F}}(x, u) \leq 1$.

The above expression of T2 FS is known as point-valued representation, and it provides a baseline for deriving other forms of representation. Another representation of T2 FS is as

$$\tilde{F} = \int_{x\in X} \int_{u\in[0,1]} \mu_{\tilde{F}}(x, u)/(x, u) \tag{2.21}$$

where $\int \int$ signifies union over all possible values of $x$ and $u$.

Fuzzy set $\tilde{F}$ is named an IT2 FS when $\mu_{\tilde{F}}(x, u) = 1, \forall x \in X$ and $\forall u \in J_x$ and it is generally represented as

$$\tilde{F} = \int_{x\in X} \int_{u\in[0,1]} 1/(x, u) \tag{2.22}$$

The FOU of $\tilde{F}$ is the 2-D support of $\mu_{\tilde{F}}(x, u)$ and characterized as

$$FOU(\tilde{F}) = \{(x, u) \in X \times [0, 1] | \mu_{\tilde{F}}(x, u) > 0\} \tag{2.23}$$

The FOU of $\tilde{F}$ bounded by UMF and LMF is described as

$$LMF(\tilde{F}) = \underline{\mu}_{\tilde{F}}(x) = \inf\{u | u \in [0, 1], \mu_{\tilde{F}}(x, u) > 0\} \tag{2.24}$$

and

$$UMF(\tilde{F}) = \overline{\mu}_{\tilde{F}}(x) = \sup\{u | u \in [0, 1], \mu_{\tilde{F}}(x, u) > 0\} \tag{2.25}$$

At each $x \in X$ of $\tilde{F}$, the primary membership $J_x$ is the interval set and mathematically represented as

$$J_x = \{u \in [0, 1] | \mu_{\tilde{F}}(x, u) > 0\} = [\underline{\mu}_{\tilde{F}}(x), \overline{\mu}_{\tilde{F}}(x)] \tag{2.26}$$

The secondary MF of $\tilde{F}$ is generally signified as $\mu_{\tilde{F}}(x)$ with function restriction $\mu_{\tilde{F}} : X \times [0, 1] \rightarrow [0, 1]$ to $x \in X$. The secondary MF of GT2 FS is T1 FS and is also called a vertical slice of $\mu_{\tilde{F}}(x, u)$.

The vertical slice representation of GT2 FS is expressed as a union of its all secondary T1 FSs and can be formulated as

$$\tilde{F} = \int_{x \in X} \mu_{\tilde{F}}(x)/x \tag{2.27}$$

where

$$\mu_{\tilde{F}}(x) = \mu_{\tilde{F}}(u/x) = \int_{\forall u \in [0,1]} f_s(u)/u. \tag{2.28}$$

The term $f_s(u)$ in Eq. (2.28) is known as secondary grade. The vertical slice representation for GT2 FS is very suitable for theoretical studies and various computations.

The embedded T2 FS of $\tilde{F}$, $\tilde{F}_e$, is represented by the MF

$$\mu_{\tilde{F}_e} : X \rightarrow [0, 1] \tag{2.29}$$

where

$$\mu_{\tilde{F}_e} = \mu_{\tilde{F}}(x, u(x)), \quad u(x) \in J_x. \tag{2.30}$$

The $\tilde{F}_e$ is generally described as

$$\tilde{F}_e = \int_{x \in X} [f_x(u)/u]/x, \quad u \in J_x \tag{2.31}$$

The embedded T2 FS representation is also called the wavy-slice representation or Mendel-John representation.

Most researchers prefer using IT2 FSs over GT2 FSs due to their relatively lower architectural intricacy and processing requirements. However, GT2 FSs are better equipped to handle noisy and highly uncertain data, making them more effective in specific scenarios requiring greater robustness. In [5, 15], Mendel demonstrated that GT2 FLSs outperform IT2 FLSs in delivering improved results. He also outlined the conceptual advancement from T1 to IT2 to GT2 FLSs, emphasizing that the processing required for GT2 FLSs can be effectively achieved using the mathematical frameworks of T1 FS and IT2 FS. The $\alpha$-plane representation theorem proposed by Liu [16] has inspired researchers in the field of GT2 FLSs to simplify and enhance their applicability [17, 18]. Numerous studies have highlighted the additional features of GT2 fuzzy sets (FSs). An innovative approach, which adaptively computes the optimal fuzzifier values for the $\alpha$-planes of GT2 FSs, is detailed in [19]. Furthermore, the concept of type-2 intuitionistic FSs and their associated learning techniques have been explored in [20–22]. In [23], Wu and Mendel provided a comprehensive exploration of Jaccard similarity measures for GT2 FSs. In [24], constrained type-2

(CT2) FSs were introduced. The concept of thick fuzzy sets (TFSs) for handling uncertainty in fuzzy modeling and computations is discussed in [25]. Additionally, [26] presents a detailed analysis of the efficiency of a novel type-reduction approach.

In recent studies, GT2 FLSs have found applications in diverse engineering fields. For instance, [27–29] demonstrate the use of GT2 FLSs in designing fuzzy controllers. In [30], GT2 FLSs are employed in humanoid robots for brain-machine interfaces, while [31] applies them to cognitive tasks for operator performance control. The detection of bearing faults using GT2 FLSs is explored in [32] and comprehensively reviewed in [33]. Noise robustness of T2 fuzzy controllers is analyzed in [34], and their application in medical diagnosis is showcased in [35]. These studies collectively highlight that GT2 FLSs offer superior capabilities in handling vagueness and uncertainties compared to IT2 FLSs.

## 2.6 Potential Applications

From the above study and literature on fuzzy systems, it is evident that higher-order fuzzy relations (Type-2 or Type-$n$) are considered effective approaches for enhancing the fuzziness of a relation. According to Hisdal [36], "greater fuzziness in descriptions implies the enhanced ability to handle imprecise information logically." In fuzzy systems theory, research on higher-type fuzzy sets (Type-3 or beyond) is sparse, offering opportunities for further exploration. In [37, 38], Rickard et al. developed the fuzzy subset hood theory for Type-$n$ fuzzy sets.

In [14, 39, 40], T3 and higher-order fuzzy sets are described as promising areas for future research in fuzzy systems theory. Castillo [39] proposed a method for determining the optimal value of n in Type-n fuzzy systems. Additionally, [41, 42] discuss the theory of hesitant fuzzy sets and provide a brief overview of Type-3 and Type-n fuzzy sets.

In [43], a general concept of the T3 FS MF is discussed. Aisbett et al. [44] proposed set theory operations for T3 and higher-level fuzzy sets. Fisher et al. [45] examined Type-$n$ fuzzy representation for spatial phenomena. Turksen [46] suggested that T3 to Type-$n$ fuzzy sets can be generated using an iterative algorithm initially developed for T2 fuzzy sets. Building on these contributions and aligning with Mendel's design approach for GT2 FLSs [5, 17], this book emphasizes the systematic design of IT3 T-S fuzzy systems using $\alpha$-plane representation theorems.

Considering the advantages associated with fuzzy-model-based control design, the application area related to aircraft/UAVs is explored. Modern aircraft is a well-known example of a highly nonlinear complex dynamical system in which there exist various kinds of uncertainty and unsteadiness during flight.

## 2.7 Summary

This chapter introduces the concept of FLS and its ability to handle uncertainties in design applications where data may be vague or incomplete. Uncertainties can be linguistic, associated with words, or random due to unpredictability. Fuzzy logic, developed by Lofti A. Zadeh, provides a framework to manage both types of uncertainties through T1 and T2 fuzzy systems. T1 FLS uses crisp membership functions, while T2 FLS, an extension of T1, provides additional flexibility by using secondary membership functions to handle higher degrees of uncertainty. Over time, fuzzy logic has been widely applied in various fields, such as control engineering, decision-making, and pattern recognition. While T1 FLS are simpler, T2 and GT2 FLS offer more robust solutions for complex, uncertain, and noisy systems. The chapter also outlines the development and potential applications of higher-order fuzzy systems, suggesting that Type-3 and higher-level fuzzy sets offer future research opportunities.

## References

1. G. J. Klir and M. J. Wierman, "Uncertainty-Based Information", Physica-Verlag, Heidelberg, 1998.
2. H. R. Berenji, "Treatment of Uncertainty in Artificial Intelligence", in H. Heer and H. Lum (eds.), *Machine Intelligence and Autonomy Aerospace Systems*, pp. 233-247, AIAA, Washington, DC, 1988.
3. J. M. Mendel, "Uncertain Rule-Based Fuzzy Logic Systems: Introduction and New Directions", Prentice-Hall, Upper Saddle River, 2001.
4. J. M. Mendel, R. I. John, and F. Liu, "Interval Type-2 Fuzzy Logic Systems Made Simple", *IEEE Trans. Fuzzy Syst.*, vol. 14, no. 6, pp. 808-821, 2006. https://doi.org/10.1109/TFUZZ.2006.879986.
5. J. M. Mendel, "General Type-2 Fuzzy Logic Systems Made Simple: A Tutorial", *IEEE Trans. Fuzzy Syst.*, vol. 22, no. 5, pp. 1162-1182, 2014. https://doi.org/10.1109/TFUZZ.2013.2286414.
6. L. A. Zadeh, "Fuzzy Sets", *Information and Control*, vol. 8, pp. 338-353, 1965.
7. R. E. Bellman and L. A. Zadeh, "Decision-Making in a Fuzzy Environment", *Management Science*, vol. 17, no. 4, pp. B141-B164, 1970.
8. R. E. Bellman and L. A. Zadeh, "Outline of a New Approach to the Analysis of Complex Systems and Decision Processes", *IEEE Trans. Syst., Man, Cybern.*, vol. SMC-3, no. 1, pp. 28-44, Jan. 1973.
9. E. H. Mamdani and S. Assilian, "An Experiment in Linguistic Synthesis with a Fuzzy Logic Controller", *Int. J. Man-Mach. Stud.*, vol. 7, no. 1, pp. 1-13, 1975.
10. L. E. Holmblad and J. J. Ostergaard, "Control of a Cement Kiln by Fuzzy Logic", in *Fuzzy Information and Decision Processes*, M. M. Gupta and E. Sanchez, Eds. North-Holland, 1981, pp. 389-399.
11. M. Sugeno, *Industrial Applications of Fuzzy Control*, Elsevier Science, 1985.
12. M. Sugeno and K. Nishida, "Fuzzy Control of a Model Car", *Fuzzy Sets Syst.*, vol. 16, no. 2, pp. 103-113, 1985.
13. K. Hirota, H. Arai, and S. Hachisu, "A Real-Time Two-Dimensional Fuzzy Robot Control System", in *Industrial Applications of Fuzzy Control*, M. Sugeno, Ed. Elsevier Science, 1989, pp. 89-95.

14. M. F. Zarandi, R. Gamasaee, and O. Castillo, "Type-1 to Type-n Fuzzy Logic and Systems", in C. Kahraman, et al. (eds.), *Fuzzy Logic in Its 50th Year*, Springer, Switzerland, 2016, pp. 129-157. https://doi.org/10.1007/978-3-319-31093-0_6.
15. J. M. Mendel, "Comparing the Performance Potentials of Interval and General Type-2 Rule-Based Fuzzy Systems in Terms of Sculpting the State Space", *IEEE Trans. Fuzzy Syst.*, vol. 27, no. 1, pp. 58-71, 2019. https://doi.org/10.1109/TFUZZ.2018.2856184.
16. F. Liu, "An Efficient Centroid Type-Reduction Strategy for General Type-2 Fuzzy Logic System", *Inf. Sci.*, vol. 178, pp. 2224-2236, 2008. https://doi.org/10.1016/j.ins.2007.11.014.
17. J. M. Mendel, F. Liu, and D. Zhai, "Alpha-Plane Representation for Type-2 Fuzzy Sets: Theory and Applications", *IEEE Trans. Fuzzy Syst.*, vol. 17, no. 5, pp. 1189-1207, 2009. https://doi.org/10.1109/TFUZZ.2009.2024411.
18. E. Ontiveros, P. Melin, and O. Castillo, "High Order Alpha-Planes Integration: A New Approach to Computational Cost Reduction of General Type-2 Fuzzy Systems", *Eng. Appl. Artif. Intell.*, vol. 74, pp. 186-197, 2018. https://doi.org/10.1016/j.engappai.2018.06.013.
19. S. Kulkarni, R. Agrawal, and F. C. H. Rhee, "Determining the Optimal Fuzzifier Range for Alpha-Planes of General Type-2 Fuzzy Sets", in *2018 IEEE International Conference on Fuzzy Systems (FUZZ-IEEE)*, pp. 1-8, Rio de Janeiro, 2018. https://doi.org/10.1109/FUZZ-IEEE.2018.8491556.
20. O. Castillo and K. Atanassov, "Comments on Fuzzy Sets, Interval Type-2 Fuzzy Sets, General Type-2 Fuzzy Sets and Intuitionistic Fuzzy Sets", in S. Melliani and O. Castillo (eds.), *Recent Advances in Intuitionistic Fuzzy Logic Systems*, Springer, Switzerland, 2019, pp. 35-43. https://doi.org/10.1007/978-3-030-02155-9_3.
21. I. Eyoh, R. John, G. De Maere, and E. Kayacan, "Hybrid Learning for Interval Type-2 Intuitionistic Fuzzy Logic Systems as Applied to Identification and Prediction Problems", *IEEE Trans. Fuzzy Syst.*, vol. 26, no. 5, pp. 2672-2685, 2018. https://doi.org/10.1109/TFUZZ.2018.2803751.
22. I. J. Eyoh, U. A. Umoh, U. G. Inyang, and J. E. Eyoh, "Derivative-Based Learning of Interval Type-2 Intuitionistic Fuzzy Logic Systems for Noisy Regression Problems", *Int. J. Fuzzy Syst.*, vol. 22, no. 3, pp. 1007-1019, 2020. https://doi.org/10.1007/s40815-020-00806-z.
23. D. Wu and J. M. Mendel, "Similarity Measures for Closed General Type-2 Fuzzy Sets: Overview, Comparisons, and a Geometric Approach", *IEEE Trans. Fuzzy Syst.*, vol. 27, no. 3, pp. 515-526, 2019. https://doi.org/10.1109/TFUZZ.2018.2862869.
24. P. D'Alterio, J. M. Garibaldi, R. I. John, and A. Pourabdollah, "Constrained Interval Type-2 Fuzzy Sets", *IEEE Trans. Fuzzy Syst.*, vol. 29, no. 5, pp. 1212-1225, 2021. https://doi.org/10.1109/TFUZZ.2020.2970911.
25. R. Boukezzoula, L. Jaulin, B. Desrochers, and D. Coquin, "Thick Fuzzy Sets (TFSs) and Their Potential Use in Uncertain Fuzzy Computations and Modeling", *IEEE Trans. Fuzzy Syst.*, 2020. https://doi.org/10.1109/TFUZZ.2020.3018550.
26. C. Chen, D. Wu, J. M. Garibaldi, R. I. John, J. Twycross, and J. M. Mendel, "A Comprehensive Study of the Efficiency of Type-Reduction Algorithms", *IEEE Trans. Fuzzy Syst.*, 2020. https://doi.org/10.1109/TFUZZ.2020.2981002.
27. A. Sakalli, T. Kumbasar, and J. M. Mendel, "Towards Systematic Design of General Type-2 Fuzzy Logic Controllers: Analysis, Interpretation and Tuning", *IEEE Trans. Fuzzy Syst.*, 2020. https://doi.org/10.1109/TFUZZ.2020.3016034.
28. J. Z. Shi, "A Fractional Order General Type-2 Fuzzy PID Controller Design Algorithm", *IEEE Access*, vol. 8, pp. 52151-52172, 2020. https://doi.org/10.1109/ACCESS.2020.2980686.
29. F. Baghbani, M. R. Akbarzadeh-T, and A. Akbarzadeh, "Indirect Adaptive Robust Mixed $H_2/H\infty$ General Type-2 Fuzzy Control of Uncertain Nonlinear Systems", *Appl. Soft Comput.*, vol. 72, pp. 392-418, 2018. https://doi.org/10.1016/j.asoc.2018.06.049.
30. J. Andreu-Perez, F. Cao, H. Hagras, and G. Yang, "A Self-Adaptive Online Brain–Machine Interface of a Humanoid Robot Through a General Type-2 Fuzzy Inference System", *IEEE Trans. Fuzzy Syst.*, vol. 26, no. 1, pp. 101-116, 2018. https://doi.org/10.1109/TFUZZ.2016.2637403.

31. L. A. Torres Salomao, M. Mahfouf, E. El-Samahy, and C. Ting, "Psychophysiologically Based Real-Time Adaptive General Type-2 Fuzzy Modeling and Self-Organizing Control of Operator's Performance Undertaking a Cognitive Task", *IEEE Trans. Fuzzy Syst.*, vol. 25, no. 1, pp. 43-57, 2017. https://doi.org/10.1109/TFUZZ.2016.2598363.
32. H. Hassani, J. Zarei, M. M. Arefi, and R. Razavi-Far, "zSlices-Based General Type-2 Fuzzy Fusion of Support Vector Machines With Application to Bearing Fault Detection", *IEEE Trans. Ind. Electron.*, vol. 64, no. 9, pp. 7210-7217, 2017. https://doi.org/10.1109/TIE.2017.2688963.
33. K. Mittal, A. Jain, K. S. Vaisla, O. Castillo, and J. Kacprzyk, "A Comprehensive Review on Type-2 Fuzzy Logic Applications: Past, Present and Future", *Eng. Appl. Artif. Intell.*, vol. 95, 2020. https://doi.org/10.1016/j.engappai.2020.103916.
34. E. Ontiveros-Robles, P. Melin, and O. Castillo, "Comparative Analysis of Noise Robustness of Type 2 Fuzzy Logic Controllers", *Kybernetika*, vol. 54, no. 1, pp. 175-201, 2018.
35. E. Ontiveros, P. Melin, and O. Castillo, "Comparative Study of Interval Type-2 and General Type-2 Fuzzy Systems in Medical Diagnosis", *Inf. Sci.*, vol. 525, pp. 37-53, 2020. https://doi.org/10.1016/j.ins.2020.03.059.
36. E. Hisdal, "The IF THEN ELSE Statement and Interval Valued Fuzzy Sets of Higher Type", *Int. J. Man-Machine Stud.*, vol. 15, pp. 385-455, 1981. https://doi.org/10.1016/S0020-7373(81)80051-X.
37. J. T. Rickard, J. Aisbett, and G. Gibbon, "Fuzzy Subsethood for Fuzzy Sets of Type-2 and Generalized Type-n", *IEEE Trans. Fuzzy Syst.*, vol. 17, no. 1, pp. 50-60, 2009. DOI: https://doi.org/10.1109/TFUZZ.2008.2006369.
38. J. T. Rickard, J. Aisbett, G. Gibbon, and D. Morgenthaler, "Fuzzy Subsethood for Type-n Fuzzy Sets", in *NAFIPS 2008 - Annual Meeting of the North American Fuzzy Information Processing Society*, pp. 1-6, 2008. https://doi.org/10.1109/NAFIPS.2008.4531276.
39. O. Castillo, "Towards Finding the Optimal n in Designing Type-n Fuzzy Systems for Particular Classes of Problems: A Review", *Appl. Comput. Math.*, vol. 17, no. 1, pp. 3-9, 2018.
40. W. Pedrycz, "Fuzzy Sets of Higher Type and Higher Order in Fuzzy Modeling", in A. Sadeghian and H. Tahayori (eds.), *Frontiers of Higher Order Fuzzy Sets*, Springer, New York, NY, 2015, pp. 31-49. https://doi.org/10.1007/978-1-4614-3442-9_3.
41. B. Zhu, Z. Xu, and M. Xia, "Dual Hesitant Fuzzy Sets", *J. Appl. Math.*, 2012, Article ID 879629. https://doi.org/10.1155/2012/879629.
42. V. Torra and Y. Narukawa, "On Hesitant Fuzzy Sets and Decision", in *2009 IEEE International Conference on Fuzzy Systems*, pp. 1378-1382, Jeju Island, 2009. https://doi.org/10.1109/FUZZY.2009.5276884.
43. I. B. Türkşen, "Type 2 Representation and Reasoning for CWW", *Fuzzy Sets Syst.*, vol. 127, no. 1, pp. 17-36, 2002. https://doi.org/10.1016/S0165-0114(01)00150-6.
44. J. Aisbett, J. T. Rickard, and D. Morgenthaler, "Intersection and Union of Type-n Fuzzy Sets", in *International Conference on Fuzzy Systems*, pp. 1-8, 2010. https://doi.org/10.1109/FUZZY.2010.5584174.
45. P. Fisher, T. Cheng, and J. Wood, "Higher Order Vagueness in Geographical Information: Empirical Geographical Population of Type-n Fuzzy Sets", *Geoinformatica*, vol. 11, no. 3, pp. 311-330, 2007. https://doi.org/10.1007/s10707-006-0009-5.
46. I. B. Türkşen, "From Type 1 to Full Type N Fuzzy System Models", *J. Mult.-Valued Logic Soft Comput.*, vol. 22, no. 4-6, pp. 543-560, 2014.

# Chapter 3
# Type-1 Fuzzy Model Based Control System

**Abstract** A design application of a Type-1 Fuzzy Model-Based (FMB) control system for generic aircraft and UAVs is presented in this chapter. A fuzzy model and a controller connected in a closed-loop configuration form the FMB control, an emerging non-linear control technique. The current theoretical framework created in this literature on fuzzy systems is examined and modified for use in aircraft-related applications. For modelling and presentation, the longitudinal aircraft dynamics in the short period mode are taken into account. The nonlinear dynamics formulae, concerning several representative locations (equilibrium points) of the flight envelope, along with a few fuzzy rules, are used to create the fuzzy model of longitudinal aircraft dynamics. While the aircraft flight envelope parameters, like operating altitude and Mach Number, are referred to as premise parameters, the consequent parameters of the fuzzy model are parts of the stability and control derivative matrix. The feedback gains of the decay rate fuzzy controller with limits on the state and input control parameters are found by solving the LMI stability requirements. The closed-loop response of the FMB controller is demonstrated under three initial flying conditions. The simulation results show that the proposed Type-1 FMB controller performs well at several indicated operating locations of the flight envelope. It stabilises the aircraft dynamics and provides better transient results.

## 3.1 Introduction

Modern aircraft can perform a wide range of flying manoeuvres in a wide range of flight conditions within their flight envelope. The aircraft dynamic behaviour is extremely complex and nonlinear due to the unsteady aerodynamics and flow-field separation exist during these manoeuvres and other flying phases [1, 2]. Nonlinearity and ambiguity are two of the many intricate considerations that need to be studied when designing and analysing a control system. The Flight Control System (FCS) is responsible for customising the aircraft's dynamic characteristics at each point of the operating envelope, as well as provides necessary level of controllability and stabilisation [2–5].

 
D. J. Singh et al., *Fuzzy Systems with Applications to Modeling and Control of an Aircraft*, Algorithms for Intelligent Systems, https://doi.org/10.1007/978-981-95-7888-7_3

Takagi-Sugeno (T-S) Fuzzy Model-Based (FMB) control is a dynamic non-linear control method of various system classes. Numerous scholars have already emphasised its advantages and possible applications in their research [6–8]. Among the advantages of T-S FMB control are: (1) the ability to interpret complex dynamics in a way that is easy for humans to understand; (2) the ability to integrate linguistic and numerical data (through human experts); (3) the ability to use simple local linear models to express complex nonlinear dynamics; and (4) the capacity to model uncertainty and ambiguity. Due to its benefits, the FMB control approach has been effectively applied in several engineering specialties [9–23].

Reference [26] offers an extensive examination of the T-S FMB control system, covering design technique, durability analysis, robustness, and implementation. References [27, 35], and [36] offer the design procedure for stabilising non-linear type frameworks. An FMB controller is developed using the Parallel Distributed Compensation (PDC) concept after the non-linear plant has been described using the T-S fuzzy form. In [28], the PDC principle is used to build decay rate-aware LMI-based methods of design. Fuzzy control design describes nonlinear structures using T-S fuzzy models and PDC algorithms.

In [29, 30], a LMI-driven design methodology for fuzzy regulators is presented using PDC and less restrictive stability criteria. In [31] presents Fuzzy Gain Scheduled (FGS) control, a gain-scheduled control technique. The model recognition method for Linear Parameter Varying (LPV) aircraft using fuzzy logic is presented and examined in [32]. In regard to the evolution of fuzzy control theory, [33] to [34] present numerous conceptual and practical investigations.

Inspired by the positive aspects of fuzzy controllers and related concepts, an application-driven study is examined for generic aeroplanes. The theoretical framework proposed by Tanaka et al. [23, 26] is taken into account for our application. The flight envelope variables, like operating altitude and Mach Number, are the premise variables of the fuzzy model, and the stability and control derivative matrices are the consequent parameters. For modeling and demonstration purposes, the short-period mode of longitudinal airplane dynamics is taken into account. The FMB controller is designed by using the aerodynamic data of the generic aircraft, which is sourced from open source [25]. In order to design the decay rate fuzzy controller with constraints on the state and control input parameters, the LMI stability requisites are established.

## 3.2 Short Period Longitudinal Aircraft Dynamics

In this section, aircraft and combat UAVs are longitudinally controlled using the FMB method over a large flight envelope structure. For purposes of illustration, short-period longitudinal dynamics are considered here. The short-period longitudinal dynamics largely dictate the short-term pitch reaction. Angle of attack $\alpha$ and pitch rate $q$ are the states; stabilizer deflection $\delta_{el}$ is the control input parameter. The dimensional coefficients are non-linear functions of several aerodynamic parameters, assuming the fixed configuration given as

$$mV\dot{\alpha} = -L + mg\cos\gamma + mVq \tag{3.1}$$

and

$$I_y\dot{q} = \bar{M}. \tag{3.2}$$

In this case, $m$ represents the aircraft's mass, $\gamma = \theta - \alpha$ represents the flight path angle, $g$ represents the acceleration caused by gravity, and $I_y$ represents the $y$ axis moment of inertia. The pitching moment $(\overline{M})$ and lift force $(L)$ are nonlinear functions of state and control parameters, and they can be expressed as follows:

$$L = \frac{1}{2}\rho V^2 SC_L(\alpha, \delta_{el}, \dot{\alpha}, q) \tag{3.3}$$

and

$$\overline{M} = \frac{1}{2}\rho V^2 S\overline{c}C_M(\alpha, \delta_{el}, \dot{\alpha}, q). \tag{3.4}$$

In this instance, $\rho$, $S$, and $\overline{c}$ represent the mean air density, the wing surface area, and the chord length, respectively. Elevator deflection $(delta_e l)$ is the input to the longitudinal model. The aerodynamic force and moment derivatives are represented by the parameters $C_L$ and $C_M$, accordingly.

We can generate the approximation T-S fuzzy model from the given nonlinear model of short-period longitudinal aircraft dynamics by using specific fuzzy rules to linearize the nonlinear model around several operational points of significance.

For linearization, perturbed state and input variables are introduced around trimmed equilibrium points as $\delta\,\alpha = \alpha - \alpha^e$, $\delta q = q - q^e$, and $\delta(\delta_{el}) = \delta_{el} - \delta_{el}^e$. Here, $\alpha^e$, $q^e$, and $\delta_{el}^e$ are the equilibrium values of the angle of attack, the pitch rate, and the deflection of the elevator.

Partial derivative of (3.1)–(3.2)

$$mV^e\delta\dot{\alpha} = -\frac{\partial L}{\partial\alpha}\delta\alpha - \frac{\partial L}{\partial q}\delta q - mg\sin\gamma^e\delta\gamma + mV^e\delta q - \frac{\partial L}{\partial\delta_{el}}\delta(\delta_{el}) \tag{3.5}$$

$$I_y\delta\dot{q} = \frac{\partial\overline{M}}{\partial\alpha}\delta\alpha + \frac{\partial\overline{M}}{\partial q}\delta q + \frac{\partial\overline{M}}{\partial\delta_{el}}\delta(\delta_{el}) \tag{3.6}$$

For equilibrium flight conditions $mg\ sin\gamma^e = -D$, hence (3.5) can be further rewritten as

$$mV^e\delta\dot{\alpha} = -\frac{\partial L}{\partial\alpha}\delta\alpha - \frac{\partial L}{\partial q}\delta q - D\delta\alpha + mV^e\delta q - \frac{\partial L}{\partial\delta_{el}}\delta(\delta_{el}). \tag{3.7}$$

The Eqs. (3.6) and (3.7) can be simplified as,

$$V^e \delta\dot{\alpha} = Z_\alpha \delta\alpha + Z_q q + Z_{\delta_{el}} \delta(\delta_{el}) \tag{3.8}$$

and

$$\delta\dot{q} = \overline{M}_\alpha \delta\alpha + \overline{M}_q \delta q + \overline{M}_{\delta_{el}} \delta(\delta_{el}). \tag{3.9}$$

Here, the force derivatives are

$$Z_\alpha = -\frac{1}{m}\left(D + \frac{\partial L}{\partial \alpha}\right), \quad Z_q = -\frac{1}{m}\frac{\partial L}{\partial q},$$

$$Z_{\delta_{el}} = -\frac{1}{m}\frac{\partial L}{\partial \delta_{el}}$$

and moment derivatives are

$$\bar{M}_\alpha = \frac{1}{I_y}\frac{\partial \bar{M}}{\partial \alpha}, \quad \bar{M}_q = \frac{1}{I_y}\frac{\partial \bar{M}}{\partial q}, \quad \bar{M}_{\delta_{el}} = \frac{1}{I_y}\frac{\partial \bar{M}}{\partial \delta_{el}},$$

respectively. The perturbed state and control input vectors are assigned as

$$\delta\mathbf{x} = \begin{bmatrix} \delta\alpha \\ \delta q \end{bmatrix}, \quad \delta\mathbf{u} = \left[\delta(\delta_{el})\right].$$

The linearized form of (3.8)–(3.9) is

$$\mathbf{E}\delta\dot{\mathbf{x}} = \bar{\mathbf{A}}\delta\mathbf{x} + \bar{\mathbf{B}}\delta\mathbf{u} \tag{3.10}$$

where, (3.10) the representative matrices are as

$$\mathbf{E} = \begin{bmatrix} V^e & 0 \\ 0 & 1 \end{bmatrix}, \quad \bar{\mathbf{A}} = \begin{bmatrix} Z_\alpha & Z_q \\ \bar{M}_\alpha & \bar{M}_q \end{bmatrix}, \quad \bar{\mathbf{B}} = \begin{bmatrix} Z_{\delta_{el}} \\ \bar{M}_{\delta_{el}} \end{bmatrix}.$$

Representing the linearization results in standard form, we get

$$\delta\dot{\mathbf{x}} = \mathbf{A}_{sp}\delta\mathbf{x} + \mathbf{B}_{sp}\delta\mathbf{u} \tag{3.11}$$

$$\text{where } \mathbf{A}_{sp} = \mathbf{E}^{-1}\bar{\mathbf{A}} \quad \text{and} \quad \mathbf{B}_{log} = \mathbf{E}^{-1}\bar{\mathbf{B}}.$$

The nonlinear aircraft dynamics are represented as a local linear model by equation (3.11). A fuzzy T-S model at a specific operating point of the flight envelope is another way to interpret this local linear model. The premise parameters of the fuzzy model are the altitude of the aircraft ($H$) and the Mach number ($M$). The fuzzy rule base can be developed for a variety of representative operating points of the flight envelope. The derivatives of aerodynamic force and moment, namely $Z_\alpha,\ Z_q,\ Z_{\delta_{el}}$ and $\overline{M}_\alpha,\ \overline{M}_q, \overline{M}_{\delta_{el}}$, depend on the aircraft's mass, inertia, and various other aerodynamic parameters at every sample point throughout the flight envelope. They are nonlinear functions of state and control input parameters. Throughout the flight, fuel consumption, outside disturbances, store deployments, and other factors alter the mass, inertia, and aerodynamic characteristics.

FMB control, which takes into consideration changes in the dynamic system traits, is one of the emerging nonlinear control systems that may easily handle this type of system. A novel method for creating an FMB controller for general aviation and unmanned aerial vehicle applications will be presented in the following section.

## 3.3 FMB Controllers for Generic Aircraft/UAVs

Considering the base parameters aircraft operating altitude $(H)$ and mach number $(M)$, (3.12) and (3.13) define the logic fundamental for the $i^{th}$ fuzzy model and corresponding PDC controller.

$$R_p^i : \text{If } H \text{ is } N_H^i \text{ and } M \text{ is } N_M^i \text{ THEN}$$

$$\dot{\mathbf{x}}(t) = \mathbf{A}_i\mathbf{x}(t) + \mathbf{B}_i\mathbf{u}(t) \tag{3.12}$$

$$R_c^i : \text{If } H \text{ is } N_H^i \text{ and } M \text{ is } N_M^i \text{ THEN}$$

$$\mathbf{u(t)} = -\mathbf{F_i x(t)} \tag{3.13}$$

where, $N_H^i$ and $N_M^i$ are fuzzy terms for the $i^th$ rule associated with the mach number $(M)$ and aircraft altitude $(H)$.

The control rules in the PDC design are derived from the fuzzy model's equivalent guidelines. Both the fuzzy model and the controller possess identical fuzzy sets in the premise section. Accordingly, $\mu_{N_H^i}$ and $\mu_{N_M^i}$ are the MFs for fuzzy models and fuzzy controllers that correspond to premise parameters $H$ and $M$, respectively. According to (3.14), the product norm is used to determine the membership grades for the $i^{th}$ rule as,

$$w_i(H, M) = \mu_{N_H^i} * \mu_{N_M^i} \tag{3.14}$$

The fuzzy control rule-set in state feedback is derived as

$$\mathbf{u}(t) = -\sum_{j=1}^{r} h_i(H, M)\mathbf{F}_i\mathbf{x}(t) \tag{3.15}$$

where,

$$h_i(H, M) = \frac{w_i(H, M)}{\sum_{i=1}^{r} w_i(H, M)}.$$

Here, $i = 1, 2, \ldots, r$ represent the PDC control design having fuzzy rules numbered as $r$. The closed-loop aircraft dynamics is represented as,

$$\mathbf{x}(t+1) = \sum_{i=1}^{r}\sum_{j=1}^{r} h_i(H, M)h_j(H, M)\left(\mathbf{A}_i - \mathbf{B}_i\mathbf{F}_j\right)\mathbf{x}(t) \tag{3.16}$$

The decay rate FMB control design with input and output limitations is taken into consideration in order to stabilize the short-period longitudinal dynamics. Equations (3.17)–(3.19) correspond to the LMI stability situations. By maximizing $\gamma \geq 0$, the LMI stability prerequisites meet the maximum convergence speed of the Lyapunov function, which is $\dot{V}(x) < -\gamma V(x)$. $V(x) = x^T\mathbf{P}x = x^T\mathbf{X}^{-1}x$ is a contender for the Lyapunov function, and $\mathbf{X}$ and $\mathbf{M}_i$ are the LMI variables.

$$\mathbf{X} > 0 \tag{3.17}$$

$$-\mathbf{X}\mathbf{A}_i^T - \mathbf{A}_i\mathbf{X} + \mathbf{M}_i^T\mathbf{B}_i^T + \mathbf{B}_i\mathbf{M}_i - 2\gamma\mathbf{X} > 0 \tag{3.18}$$

$$-\mathbf{X}\mathbf{A}_i^T - \mathbf{A}_i\mathbf{X} - \mathbf{X}\mathbf{A}_j^T - \mathbf{A}_j\mathbf{X} + \mathbf{M}_j^T\mathbf{B}_i^T + \mathbf{B}_i\mathbf{M}_j + \mathbf{M}_i^T\mathbf{B}_j^T + \mathbf{B}_j\mathbf{M}_i - 4\gamma\mathbf{X} > 0 \tag{3.19}$$

The state of equilibrium of closed loop dynamics (3.16) is globally asymptotically stable when LMI stability conditions (3.17)–(3.19) are feasible, and the highest possible Lyapunov function convergence speed is guaranteed by the designed FMB controller. The LMI variables in this case are $\mathbf{X} = \mathbf{P}^{-1}$ and $\mathbf{M}_i = \mathbf{F}_i\mathbf{X}$.

In the FMB controller design, in addition to the decay rate, the constraints on the control input and state are also considered. The following LMIs are used for expressing the elevator deflection constraint on the control input, or $(delta_el)$.

$$\begin{bmatrix} 1 & \mathbf{x}^T(0) \\ \mathbf{x}(0) & \mathbf{X} \end{bmatrix} \geq 0 \tag{3.20}$$

$$\begin{bmatrix} \mathbf{X} & \mathbf{M}_i^T\upsilon^T \\ \upsilon\mathbf{M}_i & \mu^2\mathbf{I} \end{bmatrix} \geq 0 \tag{3.21}$$

where $x(0)$ is the initial state of the system and parameter $\upsilon$ is unity ($\upsilon = 1$) for the single input case. For two input case parameter $\upsilon$ is denoted as $\upsilon_j$ where $j = 1, 2$ and $\upsilon_1 = [1\ 0]$, $\boldsymbol{\upsilon}_2 = [0\ 1]$. In our case, single input, i.e., elevator deflection ($\delta_{el}$) is considered and $\boldsymbol{\upsilon}$ is taken as unity. The notation $i = 1, 2, \ldots, r$ represents $r$ number of fuzzy rules and $\mu = |\delta_{el}^{max}| - |\delta_{el}^{e}|$. Here, $|\delta_{el}^{max}|$ is maximum value and $|\delta_{el}^{e}|$ is equilibrium value of elevator deflection. The LMIs represented through (3.20)–(3.21) guarantee $|\delta_{el}^{e}| \leq |\delta_{el}^{max}|$. The state constraints on angle of attack $\alpha$ and pitch rate parameters $q$ are also obtained by following LMIs.

$$\begin{bmatrix} 1 & \mathbf{x}^T(0) \\ \mathbf{x}(0) & \mathbf{X} \end{bmatrix} \geq 0 \tag{3.22}$$

$$\begin{bmatrix} \mathbf{X} & \mathbf{X}\mathbf{C}_k^T \\ \mathbf{C}_k\mathbf{X} & \lambda_k^2\mathbf{I} \end{bmatrix} \geq 0 \tag{3.23}$$

where, $k = 1, 2$ and $\lambda_1 = |\alpha^{max}| - |\alpha^e|$ and $\lambda_2 = |q^{max}| - |q^e|$. The vector

$$\mathbf{C}_1 = [1 \quad 0] \quad \text{and} \quad \mathbf{C}_2 = [0 \quad 1].$$

The FMB controllers, which satisfy LMI conditions (3.22)–(3.23), guarantee

$$|\alpha| \leq |\alpha^{\max}| \quad \text{and} \quad |q| \leq |q^{\max}|.$$

By resolving the LMI conditions stated at (3.17)–(3.23), the T-S FMB controller has been developed for generic aircraft in this chapter. The LMI solution yields positive definite matrices and feedback gain. Together with state constraints and the control input (actuator), the developed FMB controller satisfies the maximum convergence speed of the Lyapunov function.

## 3.4 Simulation Results

As shown in Table 3.1, the FMB control system is designed using the stability and control derivatives characteristics of a single-engine, jet fighter aircraft (referred to as an ECHO in the reference) under several flying environments.

A representative sample of the aircraft characteristics across the flight envelope is given by the four flight conditions listed above in Table 3.1. The following are the system and control matrices for short-term longitudinal dynamics at every equilibrium flight condition:

At Flight Condition 1 ($H = S.L$ and $M = 0.6$):

$$\mathbf{E}_1 = \begin{bmatrix} 207 & 0 \\ 0 & 1 \end{bmatrix}, \quad \bar{\mathbf{A}}_1 = \begin{bmatrix} -0.0272 & -2.605 \\ 0.0055 & -1.013 \end{bmatrix}, \quad \bar{\mathbf{B}}_1 = \begin{bmatrix} -0.721 \\ -0.364 \end{bmatrix}$$

At Flight Condition 2 ($H = 4600$ and $M = 0.8$):

$$\mathbf{E}_2 = \begin{bmatrix} 258 & 0 \\ 0 & 1 \end{bmatrix}, \quad \bar{\mathbf{A}}_2 = \begin{bmatrix} -0.023 & -1.866 \\ 0.0005 & -0.952 \end{bmatrix}, \quad \bar{\mathbf{B}}_2 = \begin{bmatrix} -0.67 \\ -0.362 \end{bmatrix}$$

**Table 3.1** Flight conditions

| Parameters | Flight conditions | | | |
|---|---|---|---|---|
| | 1 | 2 | 3 | 4 |
| Height (m) | S.L | 4600 | 9100 | 15250 |
| Mach | 0.6 | 0.8 | 0.95 | 1.7 |
| $V^e$ (m/s) | 207 | 258 | 288 | 502 |
| $Z_\alpha$ | −0.0272 | −0.023 | −0.016 | −0.008 |
| $Z_q$ | −2.605 | −1.866 | −1.5 | −0.46 |
| $Z_{\delta_{el}}$ | −0.721 | −0.67 | −0.4 | −0.4 |
| $\overline{M}_\alpha$ | 0.0055 | 0.0005 | −0.0002 | −0.0018 |
| $\overline{M}_q$ | −1.013 | −0.952 | −0.913 | −1.014 |

At Flight Condition 3 ($H = 9100$ and $M = 0.95$):

$$\mathbf{E}_3 = \begin{bmatrix} 288 & 0 \\ 0 & 1 \end{bmatrix}, \quad \bar{\mathbf{A}}_3 = \begin{bmatrix} -0.016 & -1.5 \\ -0.0002 & -0.913 \end{bmatrix}, \quad \bar{\mathbf{B}}_3 = \begin{bmatrix} -0.4 \\ -0.251 \end{bmatrix}$$

At Flight Condition 4 ($H = 15250$ and $M = 1.7$):

$$\mathbf{E}_4 = \begin{bmatrix} 502 & 0 \\ 0 & 1 \end{bmatrix}, \quad \bar{\mathbf{A}}_4 = \begin{bmatrix} -0.008 & -0.46 \\ -0.0018 & -1.014 \end{bmatrix}, \quad \bar{\mathbf{B}}_4 = \begin{bmatrix} -0.4 \\ -0.66 \end{bmatrix}$$

By allocating the proper membership function to flight envelope parameters, the fuzzy model and controller's rule base is developed from the operating flight conditions mentioned above. When designing the FMB control system, the triangular MFs are taken into account.

The fuzzy model terms $N_H^i$ for aircraft altitude and $N_M^i$ for Mach number are assigned in the following manner to create the rule base for the fuzzy model and associated controller:

Rule 1:

$$R_p^1 : \text{If } H \text{ is } N_H^1 \text{ and } M \text{ is } N_M^1 \text{ THEN } \dot{\mathbf{x}}(t) = \mathbf{A}_{sp1}\mathbf{x}(t) + \mathbf{B}_{sp1}\mathbf{u}(t) \tag{3.24}$$

$$R_c^1 : \text{If } H \text{ is } N_H^1 \text{ and } M \text{ is } N_M^1 \text{ THEN } \mathbf{u}(t) = -\mathbf{F}_1\mathbf{x}(t) \tag{3.25}$$

Rule 2:

$$R_p^2 : \text{If } H \text{ is } N_H^2 \text{ and } M \text{ is } N_M^2 \text{ THEN } \dot{\mathbf{x}}(t) = \mathbf{A}_{sp2}\mathbf{x}(t) + \mathbf{B}_{sp2}\mathbf{u}(t) \tag{3.26}$$

$$R_c^2 : \text{If } H \text{ is } N_H^2 \text{ and } M \text{ is } N_M^2 \text{ THEN } \mathbf{u}(t) = -\mathbf{F}_2\mathbf{x}(t) \tag{3.27}$$

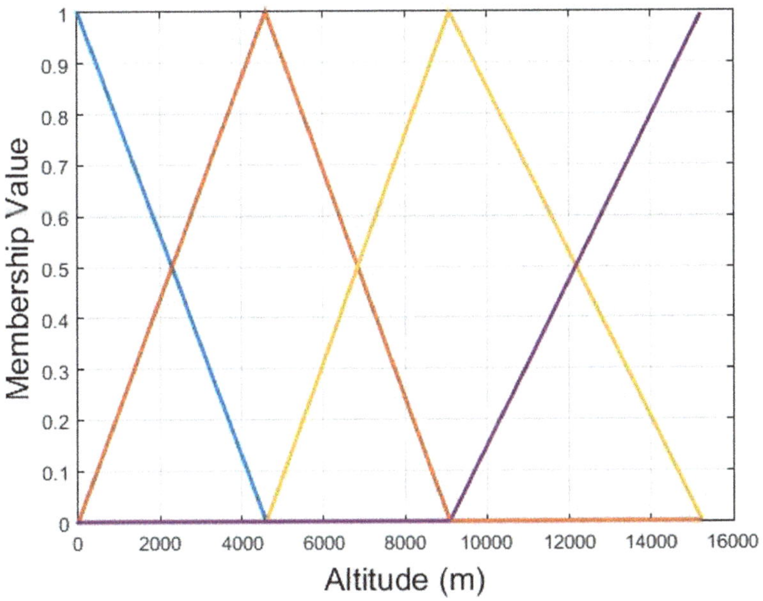


**Fig. 3.1** MF for aircraft altitude

Rule 3:

$$R_p^3 : \text{If } H \text{ is } N_H^3 \text{ and } M \text{ is } N_M^3 \text{ THEN } \dot{\mathbf{x}}(t) = \mathbf{A}_{sp3}\mathbf{x}(t) + \mathbf{B}_{sp3}\mathbf{u}(t) \tag{3.28}$$

$$R_c^3 : \text{If } H \text{ is } N_H^3 \text{ and } M \text{ is } N_M^3 \text{ THEN } \mathbf{u}(t) = -\mathbf{F}_3\mathbf{x}(t) \tag{3.29}$$

Rule 4:

$$R_p^4 : \text{If } H \text{ is } N_H^4 \text{ and } M \text{ is } N_M^4 \text{ THEN } \dot{\mathbf{x}}(t) = \mathbf{A}_{sp4}\mathbf{x}(t) + \mathbf{B}_{sp4}\mathbf{u}(t) \tag{3.30}$$

$$R_c^4 : \text{If } H \text{ is } N_H^4 \text{ and } M \text{ is } N_M^4 \text{ THEN } \mathbf{u}(t) = -\mathbf{F}_4\mathbf{x}(t) \tag{3.31}$$

The aircraft dynamic behavior in the corresponding zone of the flight envelope is described by the four rules mentioned above. The operating altitude $H$ and mach number $M$, which are the parameters of the flight envelope, are regarded as premise parameters for each zone, while the representative matrices **A** and **B** are regarded as the consequent components of the fuzzy model. These parameters change as aircraft operating zones and flight configurations change, and there is also some degree of nonlinearity and uncertainty.

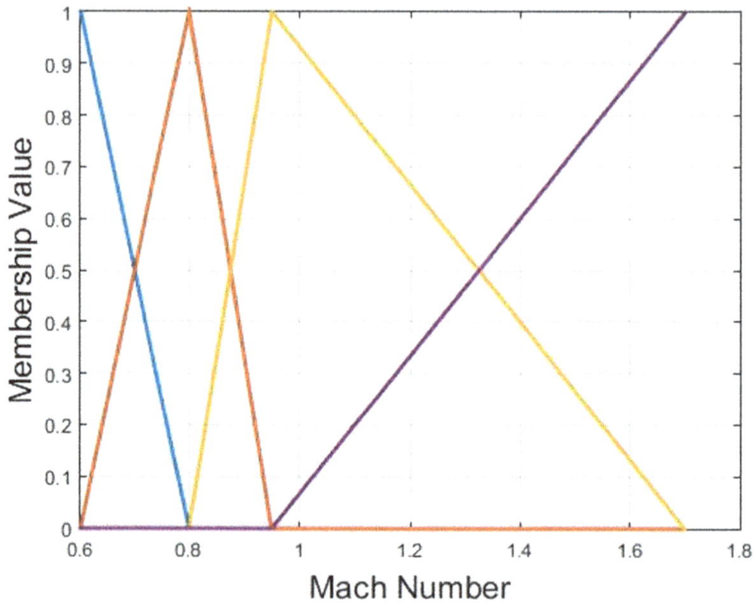


**Fig. 3.2** MF for aircraft mach number

The fuzzy model with a suitable MF (Fig. 3.1) from each fuzzy rule captures the dynamic behavior of the aircraft in each of the four zones. This indicates that the dynamic properties of the aircraft in each of the four operational zones can be sufficiently explained by a single fuzzy model that takes into consideration changes in mass, inertia, and other aerodynamic parameters. Figures 3.2 and 3.3, respectively, display the visual representation of the premise (flight envelope) parameter membership functions.

The LMI stability conditions (3.17)–(3.23) are solved to determine the controller parameters. LMI stability conditions (3.20)–(3.21) consist of the constraint on control input (elevator deflection), and LMI stability requirements (3.22)–(3.23) involve the constraint on states (angle of attack and pitch rate). In the design the constraint on control input with respect to deflection of elevator $|\delta_{el}^{max}|$ is taken as 20 deg and state constraint on $\alpha$ and $q$ are $|\alpha^{max}| \leq 65$ deg and $|q^{max}| \leq 15$ deg/s, respectively.

Thus, controller parameters obtained from the solution of LMI stability conditions (3.17)–(3.23) cater for system stability along with constraint requirements.

The positive definite matrix $\mathbf{X}$ and feedback gains $\mathbf{F}_i$ (for $i = 1, 2, 3, 4$) obtained by solving the LMI stability conditions (3.17)–(3.23) are as follows:

$$\mathbf{X} = \begin{bmatrix} 3.26 & -0.504 \\ -0.504 & 1.3284 \end{bmatrix}, \quad \mathbf{F}_1 = [0.234 \quad 8.003], \quad \mathbf{F}_2 = [0.532 \quad 7.213],$$

$$\mathbf{F}_3 = [1.721 \quad 9.321], \quad \mathbf{F}_4 = [0.852 \quad 5.797]$$

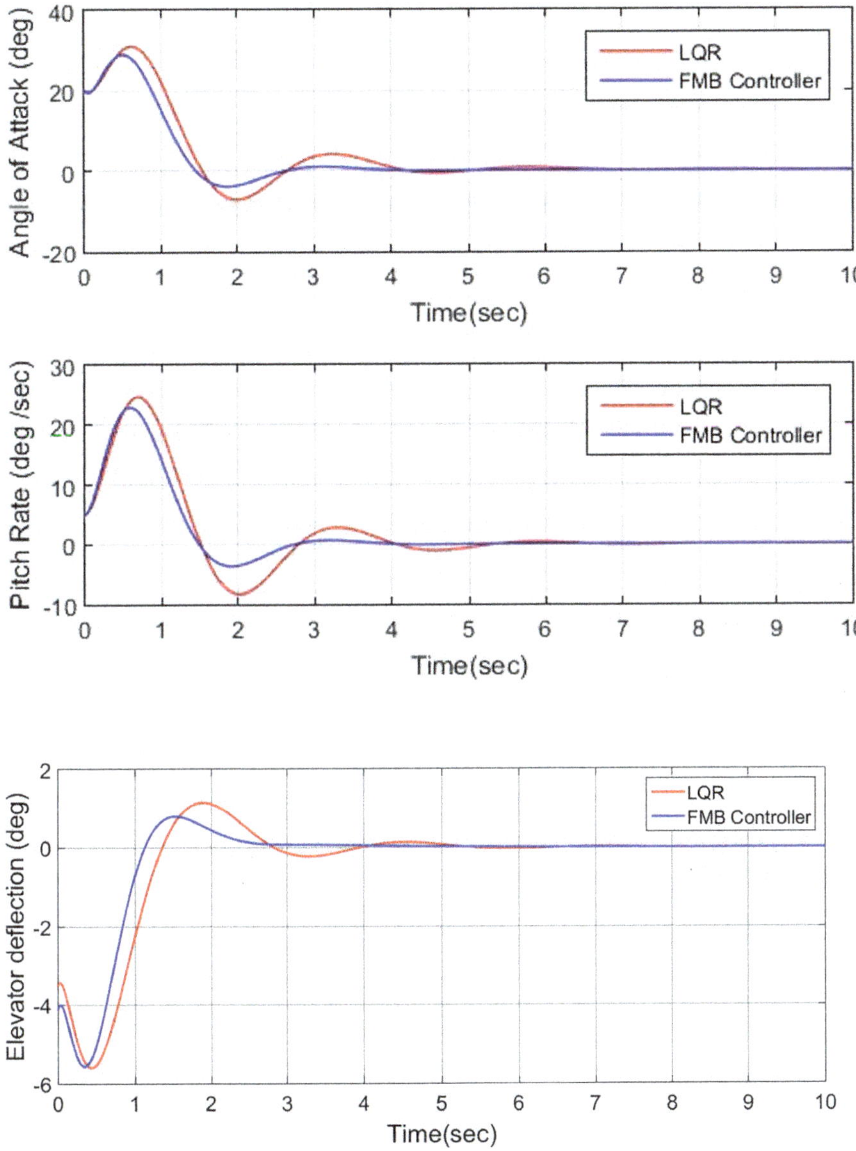


**Fig. 3.3** Initial condition 1: FMB controller performance

The focus is on the stabilisation constraint, which entails creating an FMB controller that, if we start at any random point near the operational point, should cause the closed-loop system trajectory to reach that point.

As per the desired flight conditions (flight condition 2 ($H = 4600$ and $M = 0.8$) nearby operating points), the following three sets of initial points are considered for state parameters $\alpha$ (deg) and $q$ (deg/s) to carry out the simulation studies. $(\alpha^1, q^1) \rightarrow (20, 5)$, $(\alpha^2, q^2) \rightarrow (40, 10)$, $(\alpha^3, q^3) \rightarrow (60, 15)$

Figures 3.4 and 3.6, respectively, show the closed-loop performance of the designed FMB controller from the aforementioned baseline circumstances to the equilibrium point, or $(\alpha^e, q^e) \rightarrow (0, 0)$. Equation (3.16) represents the closed-loop aircraft dynamics in the FMB controller response simulation. The simulation makes use of the controller feedback gains$\mathbf{F}_i$ for $i = 1, 2, 3, 4$ that are found by solving the LMI stability conditions (3.17)–(3.23). The normalized membership function weights $h_i(H.M)$ for the fuzzy model and $h_j(H, M)$ for the fuzzy controller are calculated as per (3.14) by taking membership values from Figs. 3.2 and 3.3 for respective flight conditions.

The FMB controller's capabilities are also compared with the linear quadratic regulator (LQR) based optimal controller. The elements of state and control weighting matrices of LQR are selected by using Bryson's principle [24] along with an iteration process during the simulation [2]. For the simulation purpose, LQR gain parameters as obtained for the linearized model corresponding to flight condition 2 ($H = 4600$ and $M = 0.8$) are utilized. The closed-loop performance of LQR in comparison with the FMB controller is presented for nearby operating points in the close vicinity of flight conditions 2 (Fig. 3.5).

The required control input, i.e., elevator deflection for the respective initial condition, is also presented in these figures. The maximum value of elevator deflection is $-5$, $-12$, and $-15$ deg with respective initial conditions. The negative sign denotes the upward deflection of the elevator trailing edge, which results in the pitching up moment.

The simulation results demonstrate the effectiveness of the proposed FMB controller at several specified operating areas of the flight envelope. In addition to stabilizing the aircraft dynamics, it provides improved transient efficiency.

In contrast to traditional LQR-based controllers, the suggested FMB flight control system has a few special benefits. For instance, the effectiveness of LQR-based controllers is comparable to that of FMB controllers if we take into account specific flight conditions, such as flight condition 2 ($H = 4600$ and $M = 0.8$). However, as shown in Fig. 3.6, the FMB controller consistently offers better transient performance when we deviate slightly from the selected operating point, such as $H = 4650$ and $M = 0.83$. Therefore, the efficiency of the FBM controller also stays satisfactory outside of the design points (operating points determined by a set of flight conditions), demonstrating the additional robustness characteristics of FMB controllers.

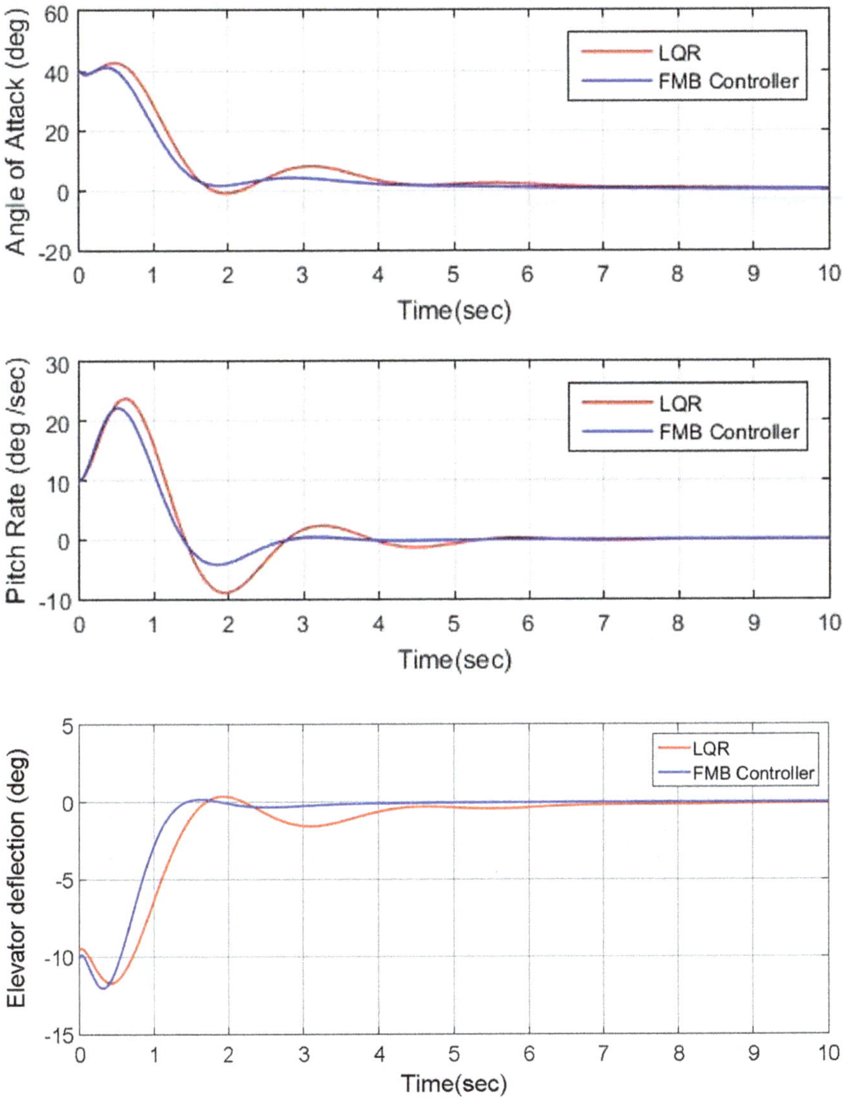


**Fig. 3.4** Initial condition 2: FMB controller performance

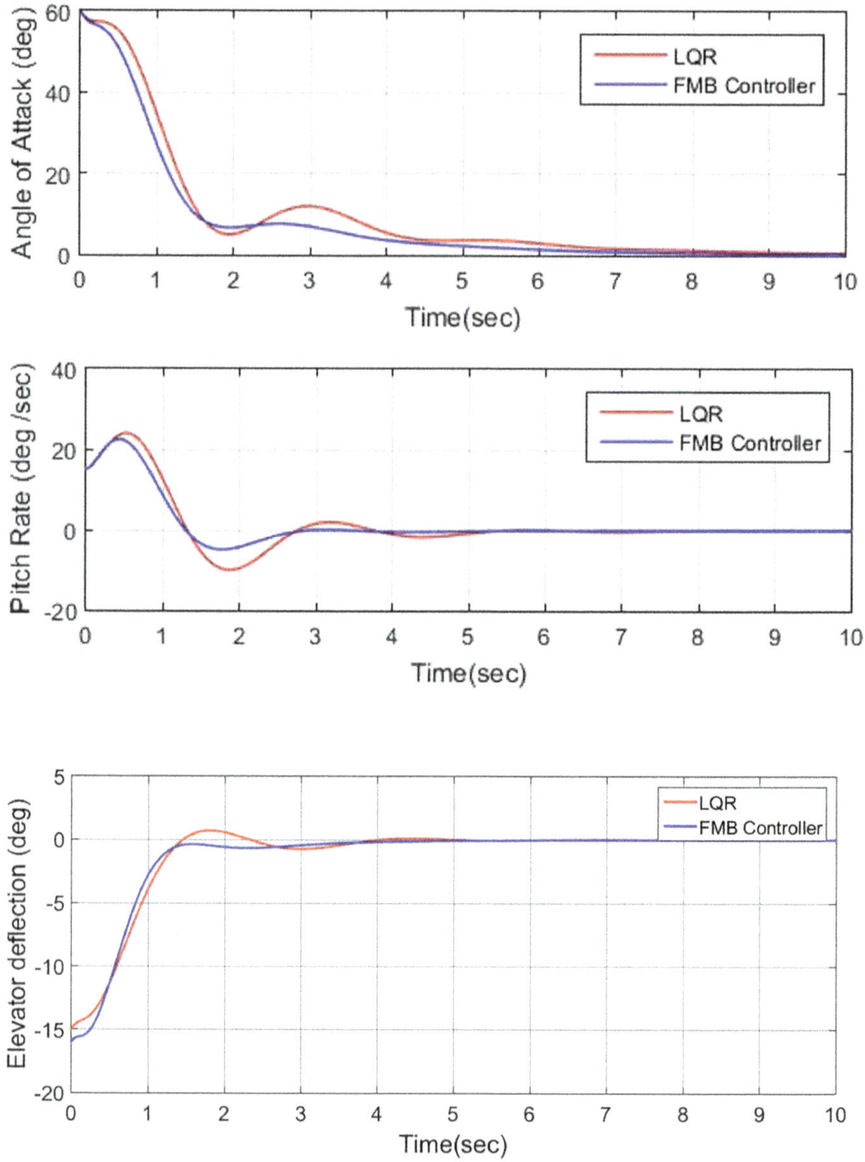


**Fig. 3.5** Initial condition 3: FMB controller performance

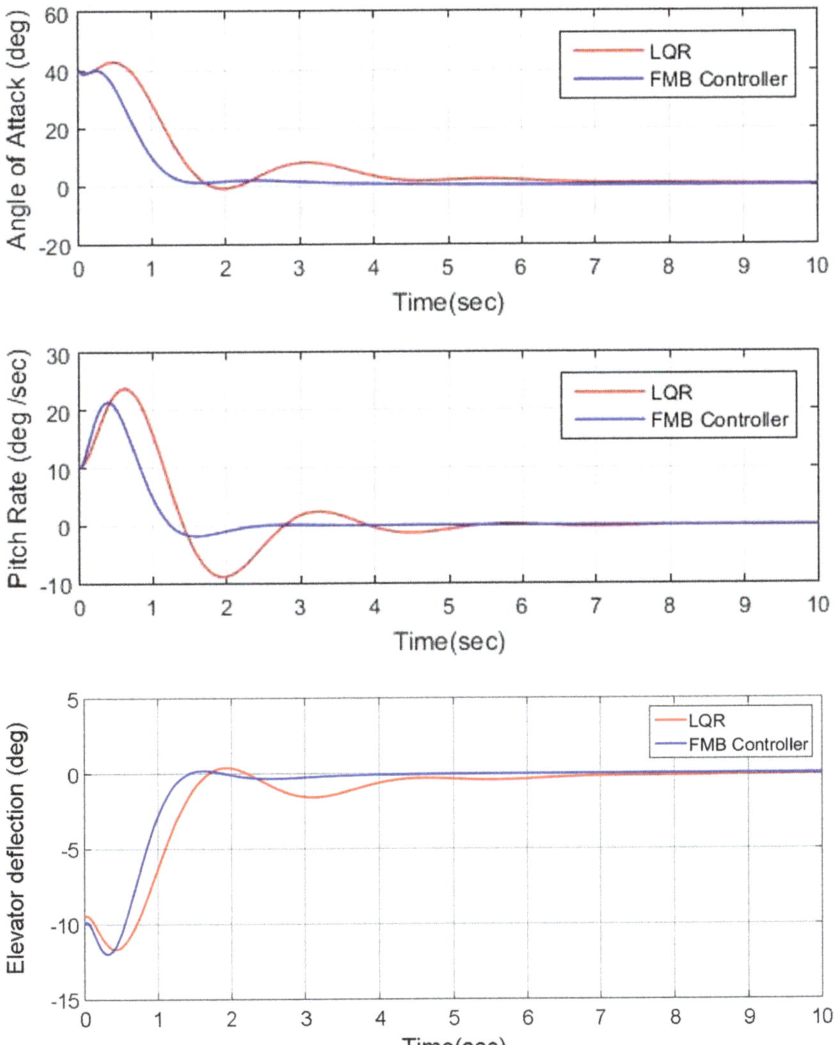


**Fig. 3.6** FMB Controller performance for arbitrary operating point (H = 4650 Km and M = 0.83)

## 3.5 Summary

The application of fuzzy control systems to UAVs and traditional aircraft in the aerospace sector is examined in this chapter. There are some unique advantages to the fuzzy control system when compared to conventional control techniques. These advantages include the capacity to incorporate both qualitative information in the form of IF-THEN fuzzy rules and quantitative information in the form of a local linear model of aircraft dynamics. Gaining knowledge of both types of aircraft dynamics aids in the design of a flight control system that can operate at peak efficiency in every operational zone of the flight envelope during different flight phases. A fuzzy controller that stabilizes the short-period mode of longitudinal aeroplane dynamics is developed for modeling and demonstration purposes. During the design stage, the decay rate controller with state and control input parameter constraints is taken into account. The LMI stability criteria must be solved in order to obtain the state feedback gains. Three initial operating points are used to illustrate the controller's closed-loop performance. According to the simulation study, the developed fuzzy controller is a good fit for a variety of applications, such as aeroplanes and unmanned aerial vehicles.

## References


1. F.R.Webster and D. Purifoy, "X-29 High Angle-of-Attack Flying Qualities." AFFTC-TR-91-15, 1991.
2. B.L. Stevens and F.L. Lewis, "Aircraft Control and Simulation," John Wiley & Sons, 2003.
3. E. Oland, R. Kristiansen and J. T. Gravdahl, "A Comparative Study of Different Control Structures for Flight Control With New Results," in IEEE Transactions on Control Systems Technology, vol. 28, no. 2, pp. 291-305, 2020, https://doi.org/10.1109/TCST.2018.2873507.
4. M. C. Palframan, J. M. Fry and M. Farhood, "Robustness Analysis of Flight Controllers for Fixed-Wing Unmanned Aircraft Systems Using Integral Quadratic Constraints," in IEEE Transactions on Control Systems Technology, vol. 27, no. 1, pp. 86-102, 2019, https://doi.org/10.1109/TCST.2017.2766598.
5. M. Mattei and V. Scordamaglia, "A Full Envelope Small Commercial Aircraft Flight Control Design Using Multivariable Proportional-Integral Control," in IEEE Transactions on Control Systems Technology, vol. 16, no. 1, pp. 169-176, 2008, https://doi.org/10.1109/TCST.2007.903098.
6. G. Feng, "A Survey on Analysis and Design of Model-Based Fuzzy Control Systems," in IEEE Transactions on Fuzzy Systems, vol. 14, no. 5, pp. 676-697, 2006, https://doi.org/10.1109/TFUZZ.2006.883415.
7. A. Nguyen, T. Taniguchi, L. Eciolaza, V. Campos, R. Palhares and M. Sugeno, "Fuzzy Control Systems: Past, Present and Future," in IEEE Computational Intelligence Magazine, vol. 14, no. 1, pp. 56-68, 2019.
8. R.-E. Precup and H. Hellendoorn, A survey on industrial applications of fuzzy control, Computers in Industry, vol. 62, no. 3, pp. 213-226, 2011.
9. K. Krishnakumar, P. Gonsalves, A, Satyadas and G. Zacharias, "Hybrid Fuzzy Logic Flight Controller Synthesis via Pilot Modeling," Journal of Guidance, Control and Dynamics, Vol 18, No. 5, Sept-Oct 1995, https://doi.org/10.2514/3.21510.
10. G. Schram, M.A. Fernández-Montesinos and H.B. Verbruggen, "Enhancing Flight Control using Fuzzy Logic," In Fuzzy algorithms for control, pp. 325-348. Springer, Dordrecht, 1999, https://doi.org/10.1007/978-94-011-4405-6_12.
11. J. Luo and E. Lan, "Fuzzy Logic Controllers for Aircraft Flight Control," In Fuzzy Logic and Intelligent Systems. International Series in Intelligent Technologies, pp. 85-124, vol 3. Springer, Dordrecht, 1995, https://doi.org/10.1007/978-0-585-28000-4_4.
12. C. Chen, W. Tan, X. Qu and H. Li, "A Fuzzy Human Pilot Model of Longitudinal Control for a Carrier Landing Task," in IEEE Transactions on Aerospace and Electronic Systems, vol. 54, no. 1, pp. 453-466, 2018, https://doi.org/10.1109/TAES.2017.2760779.

13. J. M. Giron-Sierra and G Ortega, "A survey of stability of fuzzy logic control with aerospace applications," IFAC Proceedings Volumes, Volume 35, Issue 1,2002, Pages 67-78, ISSN 1474-6670, ISBN 9783902661746, https://doi.org/10.3182/20020721-6-ES-1901.01237.
14. H. Rong, N. Sundararajan, P. Saratchandran and G. Huang, "Adaptive fuzzy fault-tolerant controller for aircraft auto landing under failures," in IEEE Transactions on Aerospace and Electronic Systems, vol. 43, no. 4, pp. 1586-1603, 2007, https://doi.org/10.1109/TAES.2007.4441760.
15. F. Santoso, M. A. Garratt and S. G. Anavatti, "State-of-the-Art Intelligent Flight Control Systems in Unmanned Aerial Vehicles," in IEEE Transactions on Automation Science and Engineering, vol. 15, no. 2, pp. 613-627, 2018, https://doi.org/10.1109/TASE.2017.2651109.
16. X. Yu, Y. Fu, P. Li and Y. Zhang, "Fault-Tolerant Aircraft Control Based on Self-Constructing Fuzzy Neural Networks and Multivariable SMC Under Actuator Faults," in IEEE Transactions on Fuzzy Systems, vol. 26, no. 4, pp. 2324-2335, 2018, https://doi.org/10.1109/TFUZZ.2017.2773422.
17. P. Hušek and K. Narenathreyas, "Aircraft longitudinal motion control based on Takagi–Sugeno fuzzy model," Applied Soft Computing, Volume 49, Pages 269-278, 2016, ISSN 1568-4946, https://doi.org/10.1016/j.asoc.2016.07.038.
18. X. Yu, Y. Fu and X. Peng, "Fuzzy Logic Aided Fault-Tolerant Control Applied to Transport Aircraft Subject to Actuator Stuck Failures," in IEEE Transactions on Fuzzy Systems, vol. 26, no. 4, pp. 2050-2061, 2018, https://doi.org/10.1109/TFUZZ.2017.2760860.
19. E. Ting, M. Ayoubi, Optimized Fuzzy-Proportional/Integral/Derivative Controller for Aircraft Pitch Control, Journal of Aerospace Information Systems, vol. 10, pp. 414-429,2013, https://doi.org/10.2514/1.I010053.
20. L. Cervantes and O. Castillo, Design of a Fuzzy System for the Longitudi-nal Control of an F-14 Airplane. In: Castillo O., Kacprzyk J., Pedrycz W. (eds) Soft Computing for Intelligent Control and Mobile Robotics. Studies in Computational Intelligence, vol 318. Springer, Berlin, Heidelberg.2010, https://doi.org/10.1007/978-3-642-15534-5_13.
21. K Nho and R K Agarwal, "Automatic Landing System Design Using Fuzzy Logic", Journal of Guidance, Control and Dynamics, Vol 23, No. 2, March-April 2000, https://doi.org/10.2514/2.4522.
22. K. Tanaka, H. Ohtake and H. O. Wang, "A practical design approach to stabilization of a 3-DOF RC helicopter," in IEEE Transactions on Control Systems Technology, vol. 12, no. 2, pp. 315-325, 2004, https://doi.org/10.1109/TCST.2003.821963.
23. M.Tanaka, K. Yamaguchi, D. Ogura, Y. J. Chen and K. Tanaka. "Nonlinear Control of F16 Aircraft via Multiple Nonlinear Model Generation for Any Trimmed Equilibriums." in International Journal of Fuzzy Systems, Vol 16, No. 2, June 2014.
24. A.E. Bryson and Y.C. Ho, "Applied Optimal Control: Optimization, Estimation, and Control," first ed., New York, NY: Taylor & Francis, 1975.
25. McLean, Donald."Automatic flight control systems (Book)" Englewood Cliffs, NJ, Prentice Hall, 1990, 606 (1990).
26. Tanaka, Kazuo, and Hua O. Wang. Fuzzy control systems design and anal-ysis: a linear matrix inequality approach. John Wiley and Sons, 2004.
27. H. O. Wang, K. Tanaka and M. F. Griffin,"An approach to fuzzy control of nonlinear systems: stability and design issues," in IEEE Transactions on Fuzzy Systems, vol. 4, no. 1, pp. 14-23, Feb. 1996, https://doi.org/10.1109/91.481841.
28. K. Tanaka, T. Ikeda and H. O. Wang, "An LMI approach to fuzzy control-ler designs based on relaxed stability conditions," Proceedings of 6th International Fuzzy Systems Conference, 1997, pp. 171-176 vol.1, https://doi.org/10.1109/FUZZY.1997.616364.
29. H. O. Wang and K. Tanaka, "An LMI-based stable fuzzy control of non-linear systems and its application to control of chaos," Proceedings of IEEE 5th International Fuzzy Systems, 1996, pp. 1433-1438 vol.2, https://doi.org/10.1109/FUZZY.1996.552386.
30. K. Tanaka and H. O. Wang, "Fuzzy regulators and fuzzy observers: a linear matrix inequality approach," Proceedings of the 36th IEEE Conference on Decision and Control, 1997, pp. 1315-1320 vol.2, https://doi.org/10.1109/CDC.1997.657640.

31. Fujimori, Atsushi, Hidekazu Tsunetomo, and Zhen-Yu Wu. "Gainscheduled control using fuzzy logic and its application to flight control." Journal of guidance, control, and dynamics 22, no. 1 (1999): 175-178.
32. Fujimori, Atsushi, and Lennart Ljung. "Model identification of linear parameter varying aircraft systems." Proceedings of the Institution of Mechanical Engineers, Part G: Journal of Aerospace Engineering 220, no. 4 (2006): 337-346.
33. K. Tanaka, T. Taniguchi and H. O. Wang, "Model-based fuzzy control of TORA system: fuzzy regulator and fuzzy observer design via LMIs that represent decay rate, disturbance rejection, robustness, optimality," 1998 IEEE International Conference on Fuzzy Systems Proceedings. IEEE World Congress on Computational Intelligence (Cat. No.98CH36228), 1998, pp. 313-318 vol.1, https://doi.org/10.1109/FUZZY.1998.687504.
34. Wang, Hua O., Jing Li, David Niemann, and Kazuo Tanaka. "TS fuzzy model with linear rule consequence and PDC controller: a universal frame work for nonlinear control systems." In Ninth IEEE International Conference on Fuzzy Systems. FUZZ-IEEE 2000 (Cat.No. 00CH37063), vol. 2, pp. 549-554. IEEE, 2000.
35. F. Santoso, M. A. Garratt and S. G. Anavatti, "State-of-the-Art Intelligent Flight Control Systems in Unmanned Aerial Vehicles," in IEEE Transactions on Automation Science and Engineering, vol. 15, no. 2,pp. 613-627, 2018, https://doi.org/10.1109/TASE.2017.2651109.
36. E. Oland, R. Kristiansen and J. T. Gravdahl, "A Comparative Study of Different Control Structures for Flight Control With New Results," in IEEE Transactions on Control Systems Technology, vol. 28, no. 2, pp.291-305, 2020, https://doi.org/10.1109/TCST.2018.2873507.

# Chapter 4
# Interval Type-2 Fuzzy Model Based Control System

**Abstract** This chapter introduces the design of an Interval Type-2 (IT2) Takagi-Sugeno (T-S) Fuzzy Model-Based (FMB) control system for aircraft. The system employs a fuzzy controller with an IT2 T-S fuzzy model operating in closed-loop configuration. The model is developed by linearizing nonlinear aircraft dynamics at selected equilibrium points in the flight envelope and applying fuzzy rules. Control and stability derivatives form the consequent parameters of the fuzzy model, while key flight envelope characteristics such as altitude and speed define its premise parameter. The longitudinal dynamics of the X-29A aircraft are used as a test case. To enhance versatility, the design incorporates Membership Function Dependent (MFD) stability analysis and considers premise mismatches. Flight envelope parameters are included in the MFD framework to capture nonlinear variations across flight conditions. The closed-loop response of the IT2 T-S FMB controller is evaluated at trim equilibrium points for four initial angle-of-attack cases. Controller performance is then compared with fuzzy LQR and conventional LQR designs. Simulation results show that the proposed IT2 T-S FMB controller effectively stabilizes the aircraft dynamics and achieves improved transient performance relative to the benchmark controllers.

## 4.1 Introduction

As an element of current investigations to deal with a high degree of uncertainty, several scholars have recently contributed to the topic of type-2 FLS and their control applications [1, 4–7]. In [7], Sarabakha et al. built and constructed the IT2 fuzzy controller for UAV. In [8], Lam and Seneviratne introduced an innovative strategy for managing synthesis and stability analysis pertaining to the IT2 FMB control system.

In [9], Amirkhani et al. developed a quadcopter flight control system based on an IT2 fuzzy cognitive map. Since then, several conceptual and practical studies have been developed concentrating on fuzzy mathematics for controller design [18–21].

 
D. J. Singh et al., *Fuzzy Systems with Applications to Modeling and Control of an Aircraft*, Algorithms for Intelligent Systems, https://doi.org/10.1007/978-981-95-7888-7_4

Type-1 T-S FMB control and Interval Type-2 (IT2) T-S FMB control are the two methods used in the T-S FMB control design. The sharpness of premise MFs made it impossible for the Type-1 T-S FMB controllers to capture the system uncertainties [16]. In order to handle the high degree of uncertainty, the Type-1 T-S FMB control concept is the basis for the IT2 T-S FMB control method. Since the premise membership function of IT2 T-S FMB controller are defined as intervals, system uncertainties can be effectively represented through their Footprints Of Uncertainty (FOUs). Subsequently, uncertainties resulting from imprecise actuation, unknown and undiscovered nonlinearity of aircraft dynamics, and fluctuation in aerodynamic characteristics with changes in flying circumstances can be easily handled by the IT2 T-S FMB flight control system.

The design process of FMB control considers a number of criteria, such as: (1) Membership Function (MF) matching strategies between fuzzy models and fuzzy controllers; (2) Lyapunov function types; (3) the use of membership functions for stability evaluation, etc. [10].

The forms of premise membership matching are identified in the literature based on the properties of MFs and the number of designated fuzzy rules used in the construction of fuzzy models and fuzzy controllers. Premise membership matches can be classified as (1) well matched, (2) somewhat matched, or (3) imperfectly matched. When designing a fuzzy model for plants and the corresponding fuzzy controller, the designer has complete control over the number of fuzzy rules and the kind of MFs in the imperfectly matched premises of these three-premise membership matching. This flexibility reduces implementation costs and structural complexity by enabling the use of fewer fuzzy rules and simpler MFs when designing fuzzy controllers [10, 11]. The IT2 T-S FMB control system stability study depends on the premise MF information, which includes its position, dimensions, and boundary. Depending on their degree of involvement, stability analysis techniques are categorized as either Membership Function Independent (MFI) or Membership Function Dependent (MFD). MFI method overlooks a portion of the system's non-linear information content. In contrast, MFD analysis captures a significant amount of system nonlinearity information since the stability criteria account for premise MF information. Apart from this, using the MFD approach, the achieved stability conditions are also more relaxed [12, 13].

A study on the IT2 T-S FMB control design with an incomplete type premise matching scenario is provided by Lam et al. [14]. They developed and analyzed the stability parameters in their study using Linear Matrix Inequality (LMI) and the MFD technique. An application-focused study of aeroplanes is inspired by the theoretical underpinnings of their research. This chapter examines a systematic method for designing an IT2 T-S FMB control system using a longitudinal dynamic model of a generic aircraft. Apart from taking into account premise-based MFD stability requirements that are not well matched, a novel approach to mapping flight envelope parameters is proposed. The flight envelope parameters are included in the MFD stability criteria in order to account for the non-linearity associated with variations in flight conditions.

For simulation, the longitudinal dynamics of the X-29A model aircraft are taken into consideration from the open source [15]. Gaussian MFs of the interval type are used for the fuzzy model, and S-shaped for the fuzzy controller in the IT2 T-S FMB controller design. The MFD method is used to solve the LMI stability criteria, yielding the slack matrices and feedback gains. To evaluate the closed-loop effectiveness, four distinct initial Angles Of Attack (AOA) values are measured around trimmed equilibrium flying conditions. Using the trimmed equilibrium flight conditions as initial states, the IT2 T-S FMB controller's performance is further compared with optimal controllers based on Linear Quadratic Regulators (LQR) and Fuzzy Linear Quadratic Regulators (Fuzzy LQR).

## 4.2 Longitudinal Aircraft Dynamics

The state-space model of longitudinal aeroplane dynamics is presented in this section. Using small perturbation theory, the nonlinear aircraft dynamics with respect to trimmed equilibrium flight conditions are linearized to produce this representation. The force and moment components acting on an aircraft during flight are described by a system of differential equations that characterize its nonlinear longitudinal dynamics. These differential equations are expressed using state parameters including aircraft speed ($V$), angle of attack ($\alpha$), pitch angle ($\theta$), and pitch rate ($q$). The nonlinear longitudinal dynamics of aircraft can be described as (4.1)–(4.4) [2].

$$m\dot{V} = -D - mg\sin\gamma + T\cos(\alpha - \epsilon) \tag{4.1}$$

$$mV\dot{\alpha} = -L + mg\cos\gamma - T\sin(\alpha - \epsilon) + mVq \tag{4.2}$$

$$\dot{\theta} = q \tag{4.3}$$

$$I_y\dot{q} = M + M_T \tag{4.4}$$

where $m$ is mass of the aircraft, $\gamma = \theta - \alpha$ is the flight path angle, $g$ is the gravitational acceleration and $I_y$ is the $y$-axis moment of inertia. The lift ($L$), drag ($D$) aerodynamic forces and pitching moment ($M$) depend nonlinearly on state and control parameters, given by:

$$L = \frac{1}{2}\rho V^2 SC_L(\alpha, \delta_{el}, \dot{\alpha}, q), \tag{4.5}$$

$$D = \frac{1}{2}\rho V^2 SC_D(\alpha, \delta_{el}), \tag{4.6}$$

$$M = \frac{1}{2}\rho V^2 S\bar{c}C_M(\alpha, \delta_{el}, \dot{\alpha}, q), \tag{4.7}$$

where $\rho$ represents air density, $S$ denotes the wing reference area, and $\bar{c}$ is the mean aerodynamic chord. The longitudinal model uses throttle setting $(\delta_{th})$ and elevator deflection $(\delta_{el})$ as inputs, while $(C_L, C_D, C_M)$ represent aerodynamic force and moment derivatives.

The angle formed by the thrust force $(T)$ and the aircraft's fixed "zero-$alpha$" line is known as the parameter $(epsilon)$. In the same direction as $(\alpha)$, this parameter is measured positively. The thrust force, which produces a moment $M_T$ about the aircraft center of gravity, is generally dependent on the throttle setting $(\delta_{th})$ and aircraft speed $(V)$.

By linearizing the non-linear model around a number of operating regions of interest using some fuzzy rules, we can derive the approximate T-S fuzzy model using the provided nonlinear model of longitudinal aircraft dynamics.

The modified state and input variables are introduced around trimmed equilibrium points to carry out the linearization as

$\delta V = V - V^e$, $\delta\alpha = \alpha - \alpha^e$, $\delta\theta = \theta - \theta^e$, $\delta q = q - q^e$, $\delta(\delta_{th}) = \delta_{th} - \delta_{th}^e$ and $\delta(\delta_{el}) = \delta_{el} - \delta_{el}^e$

Here, $V^e, \alpha^e, \theta^e, q^e, \delta_{th}^e, \delta_{el}^e$ and correspond to the matching equilibrium values for elevator deflection, throttle setting, pitch angle, pitch rate, angle of attack, and aircraft velocity. Calculating nonlinear differential equations partial derivatives, (4.1)–(4.4) about the equilibrium conditions yields (4.8)–(4.11)

$$\begin{aligned} m\,\delta\dot{V} &= -\frac{\partial D}{\partial V}\,\delta V - \frac{\partial D}{\partial \alpha}\,\delta\alpha - mg\cos\gamma^e\,\delta\gamma \\ &+ \frac{\partial T}{\partial V}\cos(\alpha^e - \epsilon)\,\delta V - T\sin(\alpha^e - \epsilon)\,\delta\alpha \\ &- \frac{\partial D}{\partial \delta_{el}}\,\delta(\delta_{el}) + \frac{\partial T}{\partial \delta_{th}}\cos(\alpha^e - \epsilon)\,\delta(\delta_{th}) \end{aligned} \tag{4.8}$$

$$\begin{aligned} mV^e\,\delta\dot{\alpha} &= -\frac{\partial L}{\partial V}\,\delta V - \frac{\partial L}{\partial \alpha}\,\delta\alpha - \frac{\partial L}{\partial q}\,\delta q \\ &- \frac{\partial L}{\partial \dot{\alpha}}\,\delta\dot{\alpha} - mg\sin\gamma^e\,\delta\gamma \\ &+ \frac{\partial T}{\partial V}\sin(\alpha^e - \epsilon)\,\delta V - T\cos(\alpha^e - \epsilon)\,\delta\alpha \\ &+ mV^e\,\delta q - \frac{\partial L}{\partial \delta_{el}}\,\delta(\delta_{el}) + \frac{\partial T}{\partial \delta_{th}}\sin(\alpha^e - \epsilon)\,\delta(\delta_{th}) \end{aligned} \tag{4.9}$$

$$\delta\dot{\theta} = \delta q \tag{4.10}$$

$$\begin{aligned} I_y\,\delta\dot{q} &= \frac{\partial M}{\partial V}\,\delta V + \frac{\partial M}{\partial \alpha}\,\delta\alpha + \frac{\partial M}{\partial q}\,\delta q \\ &\quad + \frac{\partial M}{\partial \dot{\alpha}}\,\delta\dot{\alpha} + \frac{\partial M}{\partial \delta_{el}}\,\delta(\delta_{el}) \\ &\quad + \frac{\partial M_T}{\partial V}\,\delta V + \frac{\partial M_T}{\partial \delta_{th}}\,\delta(\delta_{th}) \end{aligned} \tag{4.11}$$

Considering equilibrium flight conditions (4.12)–(4.13)

$$mg\sin\gamma^e - T\cos(\alpha^e - \epsilon) = -D \tag{4.12}$$

$$mg\cos\gamma^e - T\sin(\alpha^e - \epsilon) = L \tag{4.13}$$

The Eqs. (4.8)–(4.11) can be further rewritten as (4.14)–(4.17):

$$\begin{aligned} \delta\dot{V} &= \big(X_v + T_v\cos(\alpha^e - \epsilon)\big)\,\delta V + X_\alpha\,\delta\alpha - g\cos\gamma^e\,\delta\theta \\ &\quad + X_{\delta_{el}}\,\delta(\delta_{el}) + T_{\delta_h}\cos(\alpha^e - \epsilon)\,\delta(\delta_h) \end{aligned} \tag{4.14}$$

$$\begin{aligned} (V^e - Z_\alpha)\,\delta\dot{\alpha} &= \big(Z_v - T_v\sin(\alpha^e - \epsilon)\big)\,\delta V + Z_\alpha\,\delta\alpha - g\cos\gamma^e\,\delta\theta \\ &\quad + (V^e + Z_q)\,q + Z_{\delta_{el}}\,\delta(\delta_{el}) - T_{\delta_h}\sin(\alpha^e - \epsilon)\,\delta(\delta_h) \end{aligned} \tag{4.15}$$

$$\delta\dot{\theta} = \delta q \tag{4.16}$$

$$-M_{\dot{\alpha}}\delta\dot{\alpha} + \delta\dot{q} = (M_v + M_{Tv})\delta V + M_\alpha\delta\alpha + M_q\delta q + M_{\delta_{el}}\delta(\delta_{el}) + M_{T\delta_h}\delta(\delta_h) \tag{4.17}$$

Here, the force derivative terms are

$$\begin{aligned} X_v &= -\frac{1}{m}\frac{\partial D}{\partial V}, \quad X_\alpha = \frac{1}{m}\left(L - \frac{\partial D}{\partial \alpha}\right), \\ X_{\delta_{el}} &= -\frac{1}{m}\frac{\partial D}{\partial \delta_{el}}, \quad Z_v = -\frac{1}{m}\frac{\partial L}{\partial V}, \quad Z_\alpha = -\frac{1}{m}\left(D + \frac{\partial L}{\partial \alpha}\right), \\ Z_q &= -\frac{1}{m}\frac{\partial L}{\partial q}, \quad Z_{\dot{\alpha}} = -\frac{1}{m}\frac{\partial L}{\partial \dot{\alpha}}, \quad Z_{\delta_{el}} = -\frac{1}{m}\frac{\partial L}{\partial \delta_{el}}, \\ T_v &= \frac{1}{m}\frac{\partial T}{\partial V}, \quad T_{\delta_h} = \frac{1}{m}\frac{\partial T}{\partial \delta_h}, \end{aligned}$$

and moment derivative terms are

$$M_V = -\frac{1}{I_y}\frac{\partial M}{\partial V}, \quad M_\alpha = \frac{1}{I_y}\frac{\partial M}{\partial \alpha}, \quad M_q = \frac{1}{I_y}\frac{\partial M}{\partial q}, \quad M_{\dot{\alpha}} = \frac{1}{I_y}\frac{\partial M}{\partial \dot{\alpha}},$$
$$M_{Tv} = \frac{1}{I_y}\frac{\partial M_T}{\partial V}, \quad M_{\delta_{el}} = \frac{1}{I_y}\frac{\partial M}{\partial \delta_{el}}, \quad M_{T_{\delta_h}} = \frac{1}{I_y}\frac{\partial M_T}{\partial \delta_h}.$$

The perturbed state and control input vector are introduced as $\delta\mathbf{x} = \begin{bmatrix} \delta V \\ \delta\alpha \\ \delta\theta \\ \delta q \end{bmatrix}$ and $\delta\mathbf{u} = \begin{bmatrix} \delta(\delta_{th}) \\ \delta(\delta_{el}) \end{bmatrix}$

The linearized form of Eqs. (4.14)–(4.17) can be express in state-space form as

$$\mathbf{E}\delta\dot{\mathbf{x}} = \bar{\mathbf{A}}\delta\mathbf{x} + \bar{\mathbf{B}}\delta\mathbf{u} \tag{4.18}$$

The representative forms of the matrices $\mathbf{E}$, $\bar{\mathbf{A}}$, $\bar{\mathbf{B}}$ are given below:

$$\mathbf{E} = \begin{bmatrix} 1 & 0 & 0\,0 \\ 0 & V^e - Z_{\dot{\alpha}} & 0\,0 \\ 0 & 0 & 1\,0 \\ 0 & -M_{\dot{\alpha}} & 0\,1 \end{bmatrix},$$

$$\bar{\mathbf{A}} = \begin{bmatrix} X_V + T_V\cos(\alpha^e - \epsilon) & X_\alpha & -g\cos\gamma^e & 0 \\ Z_V - T_V\sin(\alpha^e - \epsilon) & Z_\alpha & -g\sin\gamma^e & V^e + Z_q \\ 0 & 0 & 0 & 1 \\ M_V + M_{Tv} & M_\alpha & 0 & M_q \end{bmatrix},$$

$$\bar{\mathbf{B}} = \begin{bmatrix} T_{\delta_h}\cos(\alpha^e - \epsilon) & X_{\delta_{el}} \\ -T_{\delta_h}\sin(\alpha^e - \epsilon) & Z_{\delta_{el}} \\ 0 & 0 \\ M_{T_{\delta_h}} & M_{\delta_{el}} \end{bmatrix}.$$

After standardizing the linearization results, we get

$$\delta\dot{\mathbf{x}} = \mathbf{A}_{log}\delta\mathbf{x} + \mathbf{B}_{log}\delta\mathbf{u} \tag{4.19}$$

where

$$\mathbf{A}_{log} = \mathbf{E}^{-1}\bar{\mathbf{A}}, \quad \mathbf{B}_{log} = \mathbf{E}^{-1}\bar{\mathbf{B}}.$$

Equation (4.19) gives the local linear T-S fuzzy model approximating the nonlinear longitudinal dynamics at the operating point $(\mathbf{x}^e, \mathbf{u}^e)^i$, where $\mathbf{x}^e = [V^e, \alpha^e, \theta^e, q^e]^T$

and $\mathbf{u}^e = [\delta_{th}^e, \delta_{el}^e]^T$. For $i = 1, 2 \ldots r$ corresponding to $r$ equilibrium points (equivalently fuzzy rules) (4.19) can be rewritten as:

$$\delta\dot{\mathbf{x}} = \left(\mathbf{A}_{log}\right)_i \delta\mathbf{x} + \left(\mathbf{B}_{log}\right)_i \delta\mathbf{u} \tag{4.20}$$

Therefore, a set of locally linearized T-S fuzzy models can be produced by selecting multiple trimmed equilibrium operating points on the flight envelope (under various flight conditions). The aforementioned locally linearized models' parameters, $(\mathbf{A}_{log})_i$ and $(\mathbf{B}_{log})_i$, are regarded as a consequent part of the T-S fuzzy structure, which is smoothly connected by membership functions for the premise parameters. The use of an IT2 fuzzy system as the uncertainty representation for the aerodynamic variables involved in longitudinal aircraft dynamics will be covered in the paragraph that follows.

The dynamic equations presented in (4.14)–(4.17) comprise various aerodynamic derivatives pertaining to drag force $(X_V, X_\alpha, X_{\delta_{el}})$, lift force $(Z_\alpha, Z_q, Z_{\dot{\alpha}}, Z_{\delta_{el}})$, thrust force $(T_V, T_{th})$, and pitching moment $(M_V, M_\alpha, M_q, M_{\dot{\alpha}}, M_{T_V}, M_{\delta_{el}}, M_{T_{\delta_{th}}})$. These aerodynamic derivatives rely on the aircraft's mass, inertia, and additional aerodynamic parameters at every sample point throughout the flight envelope. They are non-linear functions of state and control input parameters. During flight, fuel consumption, outside disturbances, store deployments, and other factors alter the mass, inertia, and aerodynamic characteristics.

For example, consider two representative operating regions of the flight envelope:

Zone 1: Altitude $(H_1)$ and airspeed $(V_1)$.

When the aircraft operates within this zone, its mass is denoted by $(m_1)$, the moment of inertia by $(I_{y1})$, and the state vector by $[V_1, \alpha_1, q_1, \theta_1]$. Under these conditions, one of aerodynamic derivatives pertaining to drag force will be $X_{V_1} = -\frac{1}{m_1}\frac{\partial}{\partial V_1}\left(\frac{1}{2}\rho V_1^2 S C_D(\alpha_1, \delta_{e1})\right)$, lift force will be $Z_{V_1} = -\frac{1}{m_1}\frac{\partial}{\partial V_1}\left(\frac{1}{2}\rho V_1^2 S C_L(\alpha_1, \delta_{e1}, \dot{\alpha}_1, q_1)\right)$ and pitching moment will be $M_{V_1} = -\frac{1}{I_{y1}}\frac{\partial}{\partial V_1}\left(\frac{1}{2}\rho V_1^2 S\bar{c} C_M(\alpha_1, \delta_{e1}, \dot{\alpha}_1, q_1)\right)$. Likewise, the drag, lift, and pitching moment derivatives are evaluated for Zone 1, and the corresponding matrices $\mathbf{E_1}, \bar{\mathbf{A}}_1, \bar{\mathbf{B}}_1$ are obtained.

Zone 2: Altitude $(H_2)$ and aircraft speed $(V_2)$.

When operating in this zone, the aircraft's mass is $(m_2)$, inertia $(I_{y2})$, and the state vector is $[V_2, \alpha_2, q_2, \theta_2]$. One of the aerodynamic derivatives pertaining to drag, the lift force will be $X_{V_2} = -\frac{1}{m_12}\frac{\partial}{\partial V_2}\left(\frac{1}{2}\rho V_2^2\, S C_D(\alpha_2, \delta_{e1})\right)$, $Z_{V_2} = -\frac{1}{m_2}\frac{\partial}{\partial V_2}\left(\frac{1}{2}\rho V_2^2\, S C_L(\alpha_2, \delta_{e1}, \dot{\alpha}_2, q_2)\right)$ and pitching moment will be

$M_{V_2} = -\frac{1}{I_{y2}}\frac{\partial}{\partial V_1}\left(\frac{1}{2}\rho V_2^2 S\bar{c} C_M(\alpha_2, \delta_{e1}, \dot{\alpha}_2, q_2)\right)$.

From above, It is evident that for both zones, the stability and control matrices $\mathbf{A}$log and $\mathbf{B}$log, obtained from $\mathbf{E}, \bar{\mathbf{A}}, \bar{\mathbf{B}}$, are distinct. The aircraft dynamics in each zone can be modeled using the following IF-THEN fuzzy rules:

**Rule 1**: IF the aircraft operates in Zone 1, characterized by altitude $H_1$ and speed $V_1$, THEN its dynamic behavior is represented using the matrices $\mathbf{E_1}, \bar{\mathbf{A}}_1, \bar{\mathbf{B}}_1$.

**Rule 2**: IF the aircraft operates in Zone 2, defined by altitude $H_2$ and speed $V_2$, THEN its dynamic behavior is represented by the matrices $\mathbf{E_2}$, $\bar{\mathbf{A}}_2$, $\bar{\mathbf{B}}_2$.

The dynamic behavior of the aircraft in the respective zone of the flight envelope is described by both of the aforementioned rules. The flight envelope variables, or operating altitude $H$ and aircraft speed $V$, are regarded as corresponding matrices and premise variables for each zone: $\mathbf{E}$, $\bar{\mathbf{A}}$, $\bar{\mathbf{B}}$, and are considered as consequent part of fuzzy model. The dynamic behavior of the aircraft in the respective zone of the flight envelope is described by both of the aforementioned rules. These parameters change as aircraft operating zones and flight configurations change, and there is also some degree of non-linearity and uncertainty.

By applying fuzzy blending of the flight envelope parameters with suitable membership functions from both rules, a resulting fuzzy model is obtained that captures the aircraft's dynamic behavior across both zones. In other words, the aircraft's dynamics in the two operating regions can be represented by a single fuzzy model that accounts for variations in mass $m \in [m_1, m_2]$, inertia $I_y \in [I_{y1}, I_{y2}]$, and other aerodynamic properties. The interval-based MF of flight envelope parameters in the IT2 fuzzy model is utilized for fuzzy blending of the dynamic properties of the aircraft. As a result, interval type-2 fuzzy models can readily handle parameter uncertainties brought on by changes in mass, inertia, and other aerodynamic characteristics as well as modifications in flight configurations. The section provides more information on how the resulting fuzzy model was formulated as a weighted sum of linearized aircraft dynamics.

## 4.3 Brief Review on Interval Type-2 T-S FMB Control System

An overview of the theoretical foundation of the IT2 T-S FMB control system is given in this section. A few mathematical formulas that Lam et al. proposed in [14] are repeated here for the sake of comprehensiveness.

The complex nonlinear dynamics of the plant can be represented using an IT2 T-S fuzzy model comprising $p$ IF-THEN rules, as described in [8, 14].

Rule i: IF $f_1(x(t))$ is $\tilde{M}_1^i$ and ... and $f_\psi(x(t))$ is $\tilde{M}_\psi^i$
THEN

$$\dot{\mathbf{x}}(t) = \mathbf{A}_i\mathbf{x}(t) + \mathbf{B}_i\mathbf{u}(t), \quad \mathbf{y}(t) = \mathbf{C}_i\mathbf{x}(t) \tag{4.21}$$

where, $\tilde{M}_\beta^i$ is an IT2 fuzzy term for $i$th rule related to function $f_\beta(x(t))$, $\beta = 1, 2, \ldots, \psi$, $i = 1, 2, \ldots, p$, and $\psi$ is the positive integer. The state parameters of system is $\mathbf{x}(t) \in \mathbb{R}^{n\times m}$, input parameters vector $\mathbf{u}(t) \in \mathbb{R}^m$ and output parameter vector $\mathbf{y}(t) \in \mathbb{R}^l$. The matrices $\mathbf{A}_i \in \mathbb{R}^{n\times n}$, $\mathbf{B}_i \in \mathbb{R}^{n\times m}$, and $\mathbf{C}_i \in \mathbb{R}^{l\times n}$ correspond to the system, input, and output matrices, respectively. The firing strength of the $i$th rule is an interval set and is given by:

$$\tilde{w}_i(\mathbf{x}(t)) = \left[\underline{w}_i(\mathbf{x}(t)), \overline{w}_i(\mathbf{x}(t))\right] \tag{4.22}$$

where,

$$\underline{w}_i(\mathbf{x}(t)) = \prod_{\beta=1}^{\psi} \underline{\mu}_{\tilde{M}_\beta^i}(f_\beta(\mathbf{x}(t))) \geq 0, \quad \bar{w}_i(\mathbf{x}(t)) = \prod_{\beta=1}^{\psi} \bar{\mu}_{\tilde{M}_\beta^i}(f_\beta(\mathbf{x}(t))) \geq 0$$

$$\bar{\mu}_{\tilde{M}_\beta^i}(f_\beta(\mathbf{x}(t))) \geq \underline{\mu}_{\tilde{M}_\beta^i}(f_\beta(\mathbf{x}(t))) \geq 0 \text{ and } \bar{w}_i(\mathbf{x}(t)) \geq \underline{w}_i(\mathbf{x}(t)) \geq 0, \forall j$$

where the fuzzy model's upper and lower membership grades are $\bar{w}_i(\mathbf{x}(t))$ and $\underline{w}_i(\mathbf{x}(t))$. The corresponding upper and lower membership functions are $\overline{\mu}_{\tilde{M}_\beta^i}\left(f_\beta(\mathbf{x}(t))\right)$ and $\underline{\mu}_{\tilde{M}_\beta^i}\left(f_\beta(\mathbf{x}(t))\right)$ for $\beta = 1, 2, \ldots, \psi$ respectively.

The dynamics and output of the IT2 T-S fuzzy system can be expressed as [13].

$$\dot{\mathbf{x}}(t) = \sum_{i=1}^{p} \tilde{w}_i(\mathbf{x}(t))(\mathbf{A}_i\mathbf{x}(t) + \mathbf{B}_i\mathbf{u}(t)) \tag{4.23}$$

$$\mathbf{y}(t) = \sum_{i=1}^{p} \tilde{w}_i(\mathbf{x}(t))\mathbf{C}_i\mathbf{x}(t), \tag{4.24}$$

where,

$$\sum_{i=1}^{p} \tilde{w}_i(\mathbf{x}(t)) = 1$$

The $c$-rule fuzzy controller, designed to regulate the IT2 T-S fuzzy model (4.23)–(4.24), is defined as follows:

Rule $j$: IF $g_1(\mathbf{x}(t))$ is $\tilde{N}_1^j$ and....... and $g_\Omega(\mathbf{x}(t))$ is $N_\Omega^j$

THEN

$$\mathbf{u}(t) = \mathbf{G}_j\mathbf{x}(t), \tag{4.25}$$

where, $\tilde{N}_\eta^j$ is the fuzzy term corresponding to the $j$th rule for $g_\eta(\mathbf{x}(t))$, with $j = 1, \ldots, c$ and $\eta = 1, \ldots, \Omega$, and $\mathbf{G}_j$ are the feedback gains to be estimated. The $j$th rule's firing strength is an interval set, given by:

$$\tilde{m}_j(\mathbf{x}(t)) = \left[\underline{m}_j(x(t)), \bar{m}_j(\mathbf{x}(t))\right] \tag{4.26}$$

where,

$$\underline{m}_j(\mathbf{x}(t)) = \prod_{\eta=1}^{\Omega} \underline{\mu}_{\tilde{N}_\eta^j}(g_\eta(\mathbf{x}(t))) \geq 0, \quad \bar{m}_j(\mathbf{x}(t)) = \prod_{\eta=1}^{\Omega} \bar{\mu}_{\tilde{N}_\eta^j}(g_\eta(\mathbf{x}(t))) \geq 0$$

$$\bar{\mu}_{\tilde{N}_\eta^j}(g_\eta, \mathbf{x}(t)) \geq \underline{\mu}_{\tilde{N}_\eta^j}(g_\eta, \mathbf{x}(t)) \geq 0, \quad \bar{m}_j(\mathbf{x}(t)) \geq \underline{m}_j(\mathbf{x}(t)) \geq 0, \quad \forall j$$

where $\bar{m}_j(\mathbf{x}(t))$ and $\underline{m}_j(\mathbf{x}(t))$ are normalized upper and lower membership grades for the fuzzy controller. The upper and lower membership functions are $\bar{\mu}_{\tilde{N}_\eta^j}(g_\eta, \mathbf{x}(t))$ and $\underline{\mu}_{\tilde{N}_\eta^j}(g_\eta, \mathbf{x}(t))$ for $\eta = 1, 2\ldots., \Omega$ respectively.

The fuzzy controller is expressed as follows,

$$\mathbf{u}(t) = \sum_{j=1}^{c} \tilde{m}_j(\mathbf{x}(t))\mathbf{G}_j\mathbf{x}(t) \tag{4.27}$$

where, $\sum_{j=1}^{c} \tilde{m}_j(\mathbf{x}(t)) = 1$.

Assume that the IT2 T-S fuzzy controller (4.27) and model (4.23) are connected as a closed-loop system, and that the membership functions have the following characteristics:

$$\sum_{i=1}^{p} \tilde{w}_i(\mathbf{x}(t)) = \sum_{j=1}^{c} \tilde{m}_j(\mathbf{x}(t)) = \sum_{i=1}^{p}\sum_{j=1}^{c} \tilde{w}_i(\mathbf{x}(t))\tilde{m}_j(\mathbf{x}(t)) = 1. \tag{4.28}$$

The IT2 T-S FMB controller is represented as:

$$\dot{\mathbf{x}}(t) = \sum_{i=1}^{p} \tilde{w}_i(\mathbf{x}(t)) \left( \mathbf{A}_i\mathbf{x}(t) + \mathbf{B}_i \sum_{j=1}^{c} \tilde{m}_j(\mathbf{x}(t))\mathbf{G}_j\mathbf{x}(t) \right)$$

$$= \sum_{i=1}^{p}\sum_{j=1}^{c} \tilde{w}_i(\mathbf{x}(t))\tilde{m}_j(x(t))(\mathbf{A}_i + \mathbf{B}_i\mathbf{G}_j)\mathbf{x}(t). \tag{4.29}$$

The nonlinear system dynamics in the IT2 T-S FMB control scheme can be represented by the weighted average of many local linear subsystems. The weight elements are defined by premise MFs that embed the system non-linearities. MFD-based stability evaluation can be used to incorporate the membership function data as its local boundary, global border, or as approximated discrete values using slack matrices. The global limit category contains the product of the fuzzy model and controller MFs in terms of their global limits (upper and lower). These product terms serve as constants in the evaluation of stability.

The premise Membership Function (MF) operational domain in regional MFD analysis is separated into a number of interconnected subdomains. MF's regional boundary values are used to record its information content (position and shape).

The lower and upper regional boundaries of MF vary as a result of changes in their position and form [10].

Stability requirements must be satisfied at all of the discrete values of projected membership grades, which are considered in stability analysis based on approximated membership functions. Typically, the piecewise linear [17], staircase [16] approximation, and polynomial membership function are employed. The Piecewise Linear Membership Function (PLMF) approximation is defined by sample discrete points, and two successive sample points are represented by linear interpolation [12].

Lam et al. use the Lyapunov stability theory to examine the stability of the IT2 FMB control system (4.29) in [14]. Stability analysis takes into account information about lower and upper MFs as well as sub-Footprint Of Uncertainties (FOUs). The following is a repeat of the stability analysis results that guarantee the asymptotic stability of the IT2 FMB system, as summed up in Theorem 2 of [14].

Asymptotic stability of the closed-loop IT2 FMB control system is ensured if matrices $\mathbf{M} = \mathbf{M}^T \in \mathbb{R}^{n\times n}$, $\mathbf{X} = \mathbf{X}^T \in \mathbb{R}^{n\times n}$, $N_j \in \mathbb{R}^{m\times n}$, and $\mathbf{W}_{ijl} = \mathbf{W}_{ijl}^T \in \mathbb{R}^{n\times n}$, $j = 1, 2, \ldots, c$; $i = 1, 2, \ldots, p$; $l = 1, 2, \ldots, \tau + 1$, such that the LMIs given in (4.30)–(4.33) are satisfied.

$$\mathbf{X} > 0 \tag{4.30}$$

$$\mathbf{W}_{ijl} \geq 0 \quad \forall i, j, l \tag{4.31}$$

$$\mathbf{Q}_{ij} + \mathbf{W}_{ijl} + \mathbf{M} > 0 \quad \forall i, j, l \tag{4.32}$$

$$\begin{aligned} \sum_{i=1}^{p}\sum_{j=1}^{c}\Big(& \bar{\delta}_{iji_1i_2\ldots i_nkl}\,\mathbf{Q}_{ij} \\ &- \left(\underline{\delta}_{iji_1i_2\ldots i_nkl} - \bar{\delta}_{iji_1i_2\ldots i_nkl}\right)\mathbf{W}_{ijl} \\ &+ \bar{\delta}_{iji_1i_2\ldots i_nkl}\,\mathbf{M}\Big) - \mathbf{M} < 0, \quad \forall i_1, i_2, \ldots, i_n, k, l \end{aligned} \tag{4.33}$$

here, $\bar{\delta}_{iji_1i_2\ldots i_nkl}$ and $\underline{\delta}_{iji_1i_2\ldots i_nkl}$, $i = 1, 2, \ldots, p$, $j = 1, 2, \ldots, c$, $i_1, i_2, \ldots, i_n = 1, 2$, $k = 1, 2, \ldots, q$, $l = 1, 2, \ldots, \tau + 1$, are predefined constant scalars;

$$\mathbf{Q}_{ij} = \mathbf{A}_i\mathbf{X} + \mathbf{X}\mathbf{A}_i^T + \mathbf{B}_i\mathbf{N}_j + \mathbf{N}_j^T\mathbf{B}_i^T \quad \text{for all } i \text{ and } j$$

The feedback gains are obtained as

$$\mathbf{G}_j = \mathbf{N}_j\mathbf{X}^{-1} \quad \text{for all } j.$$

LMIs are among the many optimization problems related to FMB control theory. Optimization problems with LMI constraints are commonly used to describe Semi-Definite Programming (SDP) problems. Due to the very attractive characteristics of LMIs, a variety of restrictions may be imposed in an LMI environment. [22, 23] report most of the well-known LMI characteristics, some of which are as follows: (1) Schur complement lemma; (2) Young's relation; (3) projection lemma; (4) vertex-based techniques; (5) multi-convexity; (6) congruence transformation, etc. Another useful feature of LMIs is the ability to concatenate multiple LMIs to produce a single LMI.

## 4.4 Interval Type-2 T-S FMB Control System for Generic Aircraft

For the IT2 T-S fuzzy model and fuzzy controller, the flight envelope representative parameters are regarded as premise parameters. In contrast, the **A** and **B** are recognized as consequent parameters.

The rule base is represented in the following form:

$$R_p^i : \text{If } V \text{ is } \tilde{M}_V^i \text{ and } H \text{ is } \tilde{M}_H^i \text{ THEN } \dot{x}(t) = A_i x(t) + B_i u(t) \tag{4.34}$$

$$R_c^j : \text{If } V \text{ is } \tilde{N}_V^j \text{ and } H \text{ is } \tilde{N}_H^j \text{ THEN } u(t) = G_j x(t) \tag{4.35}$$

where $\tilde{M}_V^i, \tilde{M}_H^i$ are fuzzy model terms for the $i$th rule and $\tilde{N}_V^j, \tilde{N}_H^j$ are fuzzy controller terms for the $j$th rule related with aircraft velocity $V$ and operating altitude $H$.

For the fuzzy model, the lower and upper MFs associated with $V$ and $H$ are $\left[\underline{\mu}_{\tilde{M}_V^i}(V), \bar{\mu}_{\tilde{M}_V^i}(V)\right]$ and $\left[\underline{\mu}_{\tilde{M}_H^i}(H), \bar{\mu}_{\tilde{M}_H^i}(H)\right]$ respectively.

Similarly, the fuzzy controller's lower and upper MFs for $V$ and $H$ are $\left[\underline{\mu}_{\tilde{N}_V^j}(V), \bar{\mu}_{\tilde{N}_V^j}(V)\right]$ and $\left[\underline{\mu}_{\tilde{N}_H^j}(H), \bar{\mu}_{\tilde{N}_H^j}(H)\right]$ respectively.

The $i$th rule firing strength is:

$$\tilde{w}_i(V, H) = [\underline{w}_i(V, H), \bar{w}_i(V, H)] \tag{4.36}$$

where $\underline{w}_i(V, H)$ and $\bar{w}_i(V, H)$ are lower and upper membership grades for the fuzzy model and can be obtained from the product norm as per (4.37)–(4.38):

$$\underline{w}_i(V, H) = \underline{\mu}_{\tilde{M}_V^i}(V) \cdot \underline{\mu}_{\tilde{M}_H^i}(H) \geq 0 \tag{4.37}$$

$$\bar{w}_i(V, H) = \bar{\mu}_{\tilde{M}_V^i}(V) \cdot \bar{\mu}_{\tilde{M}_H^i}(H) \geq 0 \tag{4.38}$$

In a similar manner, $j$th rule firing strength is:

$$\tilde{m}_j(V, H) = \left[\underline{m}_j(V, H), \bar{m}_j(V, H)\right] \tag{4.39}$$

where, $\underline{m}_j(V, H)$ and $[\bar{m}_j(V, H)$ represent the lower and upper membership grades of the fuzzy controller, which can be computed using the product norm as specified in (4.40)–(4.41)

$$\underline{m}_j(V, H) = \underline{\mu}_{\tilde{N}_V^j}(V) \cdot \underline{\mu}_{\tilde{N}_H^j}(H) \geq 0 \tag{4.40}$$

$$\bar{m}_j(V, H) = \bar{\mu}_{\tilde{N}_V^j}(V) \cdot \bar{\mu}_{\tilde{N}_H^j}(H) \geq 0 \tag{4.41}$$

The fuzzy model (4.23) and the controller (4.27) can now be represented in the forms shown in (4.42) and (4.43), respectively.

$$\dot{\mathbf{x}}(t) = \sum_{i=1}^{p} \tilde{w}_i(V, H)\left(\mathbf{A}_i\mathbf{x}(t) + \mathbf{B}_i\mathbf{u}(t)\right) \tag{4.42}$$

$$\mathbf{u}(t) = \sum_{j=1}^{c} \tilde{m}_j(V, H)\mathbf{G}_j\mathbf{x}(t) \tag{4.43}$$

Similarly, closed-loop IT2 T-S FMB control system (4.29) can be represented in the form of (4.44)

$$\dot{x}(t) = \sum_{i=1}^{p}\sum_{j=1}^{c} \tilde{w}_i(V, H)\tilde{m}_j(V, H)(\mathbf{A}_i + \mathbf{B}_i\mathbf{G}_j)\mathbf{x}(t) \tag{4.44}$$

$$\tilde{h}_{ij}(V, H) = \tilde{w}_i(V, H)\tilde{m}_j(V, H) \tag{4.45}$$

Taking this into account, the IT2 T-S FMB control system can be expressed in the following favorable form

$$\dot{x}(t) = \sum_{i=1}^{p}\sum_{j=1}^{c} \tilde{h}_{ij}(V, H)\left(\mathbf{A}_i + \mathbf{B}_i\mathbf{G}_j\right)\mathbf{x}(t) \tag{4.46}$$

Piecewise linear approximations of $\tilde{h}_{ij}(V, H)$ are performed for stability analysis. The original MFs are sampled into different sample points in this approximation. Original MF grades are linearly interpolated from the sample points.

Consider that operating domain $\Theta$ of $V$ and $\mathbf{H}$ are segmented into $\tilde{\lambda}$ interconnected sub-domains $\Theta_k$, such as $\Theta = \bigcup_{k=1}^{\tilde{\lambda}} \Theta_k$.

The lower and upper hybrid grade of membership $\underline{h}_{ij}(V, H)$ and $\bar{h}_{ij}(V, H)$ are defined as follows:

$$\underline{h}_{ij}(V, H) = \sum_{k=1}^{\tilde{\lambda}} \sum_{i_V=1}^{2} \sum_{i_H=1}^{2} \nu_{Vi_v k}(V)\nu_{Hi_H k}(H) \times \underline{\delta}_{iji_V i_H k}, \quad \forall i, j, k \tag{4.47}$$

$$\bar{h}_{ij}(V, H) = \sum_{k=1}^{\tilde{\lambda}} \sum_{i_V=1}^{2} \sum_{i_H=1}^{2} \nu_{Vi_V k}(V)\nu_{Hi_H k}(H) \times \bar{\delta}_{iji_V i_H k}, \quad \forall i, j, k \tag{4.48}$$

$$0 \leq \underline{h}_{ij}(V, H) \leq \bar{h}_{ij}(V, H) \leq 1 \tag{4.49}$$

$$0 \leq \underline{\delta}_{iji_V i_H k} \leq \bar{\delta}_{iji_V i_H k} \leq 1 \tag{4.50}$$

where, $\underline{\delta}_{iji_V i_H k}$ and $\bar{\delta}_{iji_V i_H k}$ are constant scalars. The linear functions $\nu_{Vi_V k}(V)$ and $\nu_{Hi_H k}(H)$ satisfy following properties when $V \in \Theta_k$ and $H \in \Theta_k$:

$$0 \leq \nu_{Vi_V k}(V) \leq 1 \quad 0 \leq \nu_{Hi_H k}(H) \leq 1 \tag{4.51}$$

$$\nu_{V1k}(V) + \nu_{V2k}(V) \leq 1 \quad \nu_{H1k}(H) + \nu_{H2k}(H) \leq 1 \tag{4.52}$$

$$\sum_{k=1}^{\tilde{\lambda}} \sum_{i_V=1}^{2} \sum_{i_H=1}^{2} \nu_{Vi_V k}(V)\nu_{Hi_H k}(H) = 1 \tag{4.53}$$

The stability conditions derived in Theorem 2 of [14] are used for deriving stability conditions for the IT2 T-S FMB control system for generic aircraft. These are as follows:

The closed-loop IT2 FMB control system for model aircraft is guaranteed to be stable (asymptotically) if there exist matrices $\mathbf{M} = \mathbf{M}^\top \in \mathbb{R}^{m\times n}$, $\mathbf{X} = \mathbf{X}^\top \in \mathbb{R}^{m\times n}$, $\mathbf{N}_j \in \mathbb{R}^{m\times n}$, $\mathbf{W}_{ij} = \mathbf{W}_{ij}^\top \in \mathbb{R}^{m\times n}$, $j = 1, 2, \ldots, c$; $i = 1, 2, \ldots, p$; such that the following LMIs are satisfied:

$$\mathbf{X} > 0 \tag{4.54}$$

$$\mathbf{W}_{ij} \geq 0 \quad \forall i, j \tag{4.55}$$

$$\mathbf{Q}_{ij} + \mathbf{W}_{ij} + \mathbf{M} > 0 \quad \forall i, j \tag{4.56}$$

$$\sum_{i=1}^{p}\sum_{j=1}^{c} (\bar{\delta}_{iji_V i_H k}\mathbf{Q}_{ij} - (\underline{\delta}_{iji_V i_H k} - \bar{\delta}_{iji_V i_H k})\mathbf{W}_{ij} + \bar{\delta}_{iji_V i_H k}\mathbf{M}) - \mathbf{M} < 0 \quad \forall i_V, i_H, k \tag{4.57}$$

here, $\bar{\delta}_{iji_V i_H k}$ and $\underline{\delta}_{iji_V i_H k}$, $i = 1, 2, \ldots, p$; $j = 1, 2, \ldots, c$; $i_v, i_\hbar = 1, 2$; $k = 1, 2, \ldots, q$ are predefined scalars;

$$\mathbf{Q}_{ij} = \mathbf{A}_i\mathbf{X} + \mathbf{X}\mathbf{A}_i^\top + \mathbf{B}_i\mathbf{N}_j + \mathbf{N}_j^\top\mathbf{B}_i^\top$$

for all $i$ and $j$, The controller feedback gains are obtained as

$$\mathbf{G}_j = \mathbf{N}_j X^{-1} \quad \text{for all } j.$$

Section 4.5 discusses the flight envelope parameters that are mapped as premise parameters of the IT2 T-S fuzzy model for X-29A research aircraft in the IT2 T-S FMB Control System Design for X-29A Model Aircraft. To accurately reflect the nonlinearity related to variation in flight conditions, the flight envelope parameters are embedded in stability conditions.

## 4.5 Simulation Results: IT2 T-S FMB Control System Design for X-29A Model Aircraft

This section presents the design algorithms of the IT2 T-S FMB control system for the X-29A research aircraft. Following [15], the IT2 FMB controller is developed to stabilize the longitudinal dynamics of the X-29A under trim equilibrium conditions, modeled using the small-perturbation hypothesis. The longitudinal model states are aircraft velocity ($V$), angle of attack ($\alpha$), pitch rate ($q$), and pitch angle ($\theta$), while the control inputs are canard deflection ($\delta_c$), symmetric flap deflection ($\delta_{sf}$), and strake flap deflection ($\delta_{stf}$). The state vector ($\mathbf{x}$) and control input vector ($\mathbf{u}$) are given as:

$$\mathbf{x} = \begin{bmatrix} V \ \alpha \ q \ \theta \end{bmatrix}^T, \quad \text{and} \quad \mathbf{u} = \begin{bmatrix} \delta_c \ \delta_{sf} \ \delta_{stf} \end{bmatrix}^T.$$

Three instances of trimmed equilibrium flight conditions offer a representative sampling to define the X-29A aircraft characteristic across its flight envelope, per ref [15]. These scenarios are shown at three distinct operating points, which are determined by the aircraft's Mach No. ($M$) and operating altitude ($H$).

For every trimmed equilibrium flight condition (operating point), the linearized system and control matrices of the longitudinal model are outlined below:

Operating point for case 1: V = 284 ft/s, H = 4000 ft

$$\mathbf{A}_1 = \begin{bmatrix} -0.04272 & -8.541 & -0.4451 & -32.16 \\ -0.0007881 & -0.5291 & 0.9896 & 0 \\ 0.0004010 & 3.542 & -0.2228 & 0 \\ 0 & 0 & 1 & 0 \end{bmatrix}, \quad \mathbf{B}_1 = \begin{bmatrix} -0.03385 & -0.09386 & 0.004888 \\ -0.001028 & -0.001297 & -0.0004054 \\ 0.02718 & -0.005744 & 0.01351 \\ 0 & 0 & 0 \end{bmatrix}.$$

Operating point for case 2: V = 726 ft/s, H = 20,000 ft

$$\mathbf{A}_2 = \begin{bmatrix} -0.0117 & -6.05 & -0.3139 & -32.11 \\ -0.00014 & -0.8167 & 0.994 & 0 \\ 0.0003213 & 12.14 & -0.4136 & 0 \\ 0 & 0 & 1 & 0 \end{bmatrix}, \quad \mathbf{B}_2 = \begin{bmatrix} -0.0605 & -0.158 & 0.01338 \\ -0.00088 & -0.0036 & -0.0005869 \\ 0.1345 & -0.08383 & -0.04689 \\ 0 & 0 & 0 \end{bmatrix}.$$

Operating point for case 3: V=977 ft/s, H=8000 ft

$$\mathbf{A}_3 = \begin{bmatrix} -0.05329 & -12.9 & -0.4938 & -32.15 \\ -0.0001712 & -2.241 & 0.9897 & 0 \\ -0.0009569 & 44.74 & -0.9024 & 0 \\ 0 & 0 & 1 & 0 \end{bmatrix}, \quad \mathbf{B}_3 = \begin{bmatrix} -0.2138 & 0.1213 & 0.0078 \\ -0.002852 & -0.0060 & -0.0009029 \\ 0.4720 & -0.2499 & -0.07966 \\ 0 & 0 & 0 \end{bmatrix}.$$

By allocating the appropriate membership function with the flight envelope parameters, the IT2 T-S fuzzy model and controller's rule base is imitated from the aforementioned operating conditions. The operating altitude and aircraft velocity are assigned as premise parameters for the fuzzy model and controller. For the T-S fuzzy model, the interval type Gaussian MF is taken into consideration, and for the T-S fuzzy controller, the interval type S-shaped MF is chosen.

The S-shaped and Gaussian MFs are chosen due to their clear notation and smoothness. The benefit of these curves is that they are perpetually non-zero and smooth. In addition, just two parameters are needed to generate the shapes of the Gaussian and S-shaped MFs. Three MFs are chosen for the fuzzy model because there are three instances of trimmed equilibrium flight conditions. We used the case of imperfectly matched premises in our design [11], which gives us the freedom to choose fewer rules for the fuzzy controller design (in contrast to a fuzzy model). Two MFs are chosen for the fuzzy controller design based on this.

The rule base is designed by assigning term $\tilde{M}_V^1, \tilde{M}_V^2, \tilde{M}_V^3$ for aircraft velocity $(V)$ and $\tilde{M}_H^1, \tilde{M}_H^2, \tilde{M}_H^3$ for operating altitude $(H)$ as follows:

$$R_{p_1} : \text{If } V \text{ is } \tilde{M}_V^1 \text{ and } H \text{ is } \tilde{M}_H^1 \text{ Then } \dot{\mathbf{x}}(t) = \mathbf{A}_1\mathbf{x}(t) + \mathbf{B}_1\mathbf{u}(t) \tag{4.58}$$

$$R_{p_2} : \text{If } V \text{ is } \tilde{M}_V^2 \text{ and } H \text{ is } \tilde{M}_H^2 \text{ Then } \dot{\mathbf{x}}(t) = \mathbf{A}_2\mathbf{x}(t) + \mathbf{B}_2\mathbf{u}(t) \tag{4.59}$$

$$R_{p_3} : \text{If } V \text{ is } \tilde{M}_V^3 \text{ and } H \text{ is } \tilde{M}_H^3 \text{ Then } \dot{\mathbf{x}}(t) = \mathbf{A}_3\mathbf{x}(t) + \mathbf{B}_3\mathbf{u}(t) \tag{4.60}$$

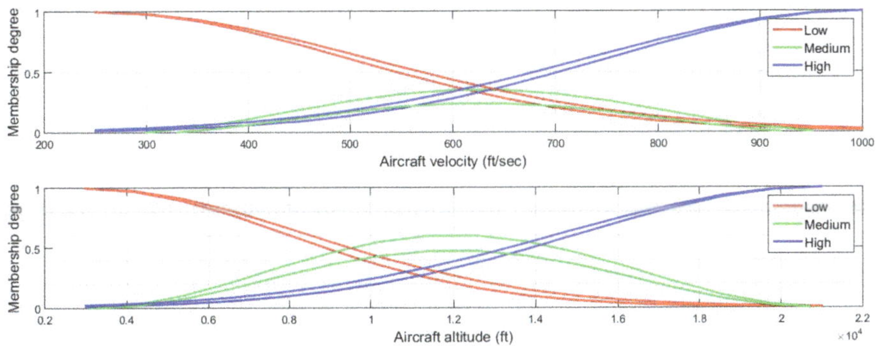


**Fig. 4.1** Premise MF for fuzzy model

The lower and upper MF corresponding to $(V)$ and $(H)$ are $\left[\mu_{\tilde{M}_V^i}(V), \bar{\mu}_{\tilde{M}_V^i}(V)\right]$ and $\left[\mu_{\tilde{M}_H^i}(H), \bar{\mu}_{\tilde{M}_H^i}(H)\right]$ (as shown in Table 4.1 and Fig. 4.1).

Two fuzzy rules with interval type S-shaped membership functions are adopted for the fuzzy controller, considering the imperfect premise matching technique in the IT2 T-S FMB design. The rule base is constructed by assigning interval type fuzzy terms $\tilde{N}_V^1, \tilde{N}_V^2$ for aircraft velocity $V$ and $\tilde{N}_H^1, \tilde{N}_H^2$ for altitude $H$. The fuzzy controller rule base is given in (4.61)–(4.62):

$$R_{c_1}: \text{If } V \text{ is } \tilde{N}_V^1 \text{ and } H \text{ is } \tilde{N}_H^1 \text{ Then } \mathbf{u}(t) = \mathbf{G}_1\mathbf{x}(t) \tag{4.61}$$

$$R_{c_2}: \text{If } V \text{ is } \tilde{N}_V^2 \text{ and } H \text{ is } \tilde{N}_H^2 \text{ Then } \mathbf{u}(t) = \mathbf{G}_2\mathbf{x}(t) \tag{4.62}$$

In the IT2 T-S fuzzy controller, the premise parameters of the lower and upper membership functions, i.e., aircraft velocity $\left(\underline{\mu}_{\tilde{N}_V}, \overline{\mu}_{\tilde{N}_V}\right)$ and operating altitude $\left(\underline{\mu}_{\tilde{N}_H}, \overline{\mu}_{\tilde{N}_H}\right)$ divide the operating regime into two portions (as shown in Table 4.2 and Fig. 4.2).

The linear (piecewise) approximation of fuzzy models and controller premise MFs are taken into consideration in order to accommodate premise MFs information during stability analysis. Flight envelope parameters, such as aircraft velocity and operating altitude under stability conditions, were highlighted by this consideration. As a result, during stability evaluation, the impact of nonlinearity resulting from variations in flight conditions is modeled. This is accomplished by mapping the flight envelope parameters' operating realm for $V \in [250\ 1000]$ and $H \in [3000\ 21000]$. The linear functions $\nu_{V1K}(V)$, $\nu_{V2K}(V)$ and $\nu_{H1K}(H)$, $\nu_{H2K}(H)$satisfying the properties of (4.51), (4.52) are chosen as (4.63) and (4.64)

$$\nu_{V1k}(V) = \left(\frac{V_{\max,k} - V}{V_{\max,k} - V_{\min,k}}\right), \quad \nu_{V2k}(V) = 1 - \nu_{V1k}(V) \tag{4.63}$$

**Table 4.1** Fuzzy model membership function representation

| *Premise parameter aircraft velocity V (ft/s)* | |
|---|---|
| Linguistic interpretation | Lower and upper MF |
| Low | $\underline{\mu}_{\tilde{M}_V} = \exp\left(-\frac{(V-250)^2}{250^2}\right), \overline{\mu}_{\tilde{M}_V} = \exp\left(-\frac{(V-250)^2}{270^2}\right)$ |
| Medium | $\underline{\mu}_{\tilde{M}_V} = 1 - \underline{\mu}_{\tilde{M}} - \underline{\mu}_{\tilde{M}_V}$ $\overline{\mu}_{\tilde{M}_V} = 1 - \overline{\mu}_{\tilde{M}} - \overline{\mu}_{\tilde{M}_V}$ |
| High | $\underline{\mu}_{\tilde{M}_V} = \exp\left(-\frac{(V-1000)^2}{250^2}\right), \overline{\mu}_{\tilde{M}_V} = \exp\left(-\frac{(V-1000)^2}{270^2}\right)$ |
| *Premise parameter aircraft altitude H (ft)* | |
| Low | $\underline{\mu}_{\tilde{M}_H} = \exp\left(-\frac{(H-3000)^2}{5000^2}\right), \overline{\mu}_{\tilde{M}_H} = \exp\left(-\frac{(H-3000)^2}{5500^2}\right)$ |
| Medium | $\underline{\mu}_{\tilde{M}_H} = 1 - \underline{\mu}_{\tilde{M}} - \underline{\mu}_{\tilde{M}_H}$ $\overline{\mu}_{\tilde{M}_H} = 1 - \overline{\mu}_{\tilde{M}} - \overline{\mu}_{\tilde{M}_H}$ |
| High | $\underline{\mu}_{\tilde{M}_H} = \exp\left(-\frac{(H-21000)^2}{5000^2}\right), \overline{\mu}_{\tilde{M}_H} = \exp\left(-\frac{(H-21000)^2}{5500^2}\right)$ |

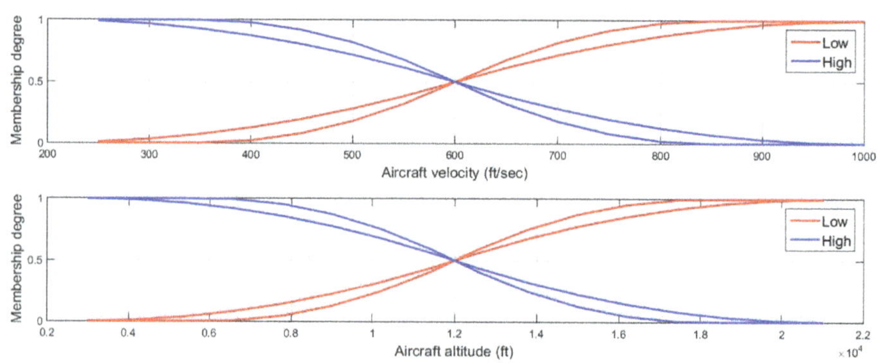


**Fig. 4.2** Fuzzy controller premise MF

$$\nu_{H1k}(H) = \left(\frac{H_{\max,k} - H}{H_{\max,k} - H_{\min,k}}\right), \quad \nu_{H2k}(H) = 1 - \nu_{H1k}(H) \tag{4.64}$$

Here, $(V_{\max,k}, V_{\min,k})$ and $(H_{\max,k}, H_{\min,k})$ represent the maximum and minimum values of $(V, H)$ in the $k$th region, with $k = 1, 2, \ldots, 15$. To adapt the stability conditions from (4.56), the constants $\delta_{ij}$ are derived from the lower membership grades of the fuzzy model ($w_i$) and fuzzy controller ($m_j$). This formulation is given in (4.65)–(4.68).

**Table 4.2** Fuzzy controller MF representation

| *Premise parameter aircraft velocity V (ft/s)* | |
|---|---|
| Linguistic interpretation | Lower and upper MF |
| Low | $\overline{\mu}_{\tilde{N}_V} = 0$ for V $\leq 200$, $\underline{\mu}_{\tilde{N}_V} = 0$ for V $\leq 250$<br>$\overline{\mu}_{\tilde{N}_V} = 2\left(\frac{V-200}{800}\right)^2$ for $200 <$ V $\leq 600$<br>$\underline{\mu}_{\tilde{N}_V} = 2\left(\frac{V-250}{700}\right)^2$ for $250 \leq$ V $\leq 600$<br>$\overline{\mu}_{\tilde{N}_V} = 1 - 2\left(\frac{V-1000}{800}\right)^2$ for $600 <$ V $\leq 1000$<br>$\underline{\mu}_{\tilde{N}_V} = 1 - 2\left(\frac{V-950}{700}\right)^2$ for $600 <$ V $\leq 950$<br>$\overline{\mu}_{\tilde{N}_V} = 1$ for V $\geq 1000$, $\underline{\mu}_{\tilde{N}_V} = 1$ for V $\geq 950$ |
| High | $\underline{\mu}_{\tilde{N}_V} = 1 - \underline{\mu}_{\tilde{N}_V}$<br>$\overline{\mu}_{\tilde{N}_V} = 1 - \overline{\mu}_{\tilde{N}_V}$ |
| *Premise parameter aircraft altitude H (ft)* | |
| Low | $\overline{\mu}_{\tilde{N}_H} = 0$ for H $\leq 3000$<br>$\underline{\mu}_{\tilde{N}_H} = 0$ for H $\leq 4000$<br>$\overline{\mu}_{\tilde{N}_H} = 2\left(\frac{H-3000}{18000}\right)^2$ for $3000 <$ H $\leq 12500$<br>$\underline{\mu}_{\tilde{N}_H} = 2\left(\frac{H-4000}{16000}\right)^2$ for $4000 <$ H $\leq 12000$<br>$\overline{\mu}_{\tilde{N}_H} = 1 - 2\left(\frac{H-3000}{18000}\right)^2$ for $12500 <$ H $\leq$ 21000<br>$\underline{\mu}_{\tilde{N}_H} = 1 - 2\left(\frac{H-4000}{16000}\right)^2$ for $12000 <$ H $\leq$ 20000<br>$\overline{\mu}_{\tilde{N}_H} = 1$ for H $\geq 21000$, $\underline{\mu}_{\tilde{N}_H} = 1$ for H $\geq$ 20000 |
| High | $\underline{\mu}_{\tilde{N}_H} = 1 - \underline{\mu}_{\tilde{N}_H}$<br>$\overline{\mu}_{\tilde{N}_H} = 1 - \overline{\mu}_{\tilde{N}_H}$ |

$$\underline{\delta}_{ijV_{\min,k}H_{\min,k}} = \underline{w}_i(V_{\min,k}, H_{\min,k})\,\underline{m}_j(V_{\min,k}, H_{\min,k}) \tag{4.65}$$

$$\underline{\delta}_{ijV_{\min,k}H_{\max,k}} = \underline{w}_i(V_{\min,k}, H_{\max,k})\,\underline{m}_j(V_{\min,k}, H_{\max,k}) \tag{4.66}$$

$$\underline{\delta}_{ijV_{\max,k}H_{\min,k}} = \underline{w}_i(V_{\max,k}, H_{\min,k})\,\underline{m}_j(V_{\max,k}, H_{\min,k}) \tag{4.67}$$

$$\underline{\delta}_{ijV_{\max,k}H_{\max,k}} = \underline{w}_i(V_{\max,k}, H_{\max,k})\,\underline{m}_j(V_{\max,k}, H_{\max,k}) \tag{4.68}$$

Similarly, $\bar{\delta}_{ij}$ are estimated from $(\bar{w}_i)$ and $(\bar{m}_i)$ using (4.69)–(4.72).

$$\bar{\delta}_{ijV_{\min,k}H_{\min,k}} = \bar{w}_i(V_{\min,k}, H_{\min,k})\,\bar{m}_j(V_{\min,k}, H_{\min,k}) \tag{4.69}$$

$$\bar{\delta}_{ijV_{\min,k}H_{\max,k}} = \bar{w}_i(V_{\min,k}, H_{\max,k})\,\bar{m}_j(V_{\min,k}, H_{\max,k}) \tag{4.70}$$

$$\bar{\delta}_{ijV_{\max,k}H_{\min,k}} = \bar{w}_i(V_{\max,k}, H_{\min,k})\,\bar{m}_j(V_{\max,k}, H_{\min,k}) \tag{4.71}$$

$$\bar{\delta}_{ijV_{\max,k}H_{\max,k}} = \bar{w}_i(V_{\max,k}, H_{\max,k})\,\bar{m}_j(V_{\max,k}, H_{\max,k}) \tag{4.72}$$

Using this adjustment, the stability condition in (4.57) is adapted to include the flight envelope parameters (aircraft speed and altitude) as premise variables of the IT2 T-S fuzzy model, capturing nonlinear effects due to changing flight conditions. The modified form is shown in (4.73)–(4.76).

$$\sum_{i=1}^{p}\sum_{j=1}^{c}\Bigg(\bar{\delta}_{ijV_{\min,k}H_{\min,k}k}\,\mathbf{Q}_{ij} - \left(\underline{\delta}_{ijV_{\min,k}H_{\min,k}k} - \bar{\delta}_{ijV_{\min,k}H_{\min,k}k}\right)\mathbf{W}_{ij} + \bar{\delta}_{ijV_{\min,k}H_{\min,k}k}\,\mathbf{M}\Bigg) - \mathbf{M} < 0 \tag{4.73}$$

$$\sum_{i=1}^{p}\sum_{j=1}^{c}\Bigg(\bar{\delta}_{ijV_{\min,k}H_{\max,k}k}\,\mathbf{Q}_{ij} - \left(\underline{\delta}_{ijV_{\min,k}H_{\max,k}k} - \bar{\delta}_{ijV_{\min,k}H_{\max,k}k}\right)\mathbf{W}_{ij} + \bar{\delta}_{ijV_{\min,k}H_{\max,k}k}\,\mathbf{M}\Bigg) - \mathbf{M} < 0 \tag{4.74}$$

$$\sum_{i=1}^{p}\sum_{j=1}^{c}\Bigg(\bar{\delta}_{ijV_{\max,k}H_{\min,k}k}\,\mathbf{Q}_{ij} - \left(\underline{\delta}_{ijV_{\max,k}H_{\min,k}k} - \bar{\delta}_{ijV_{\max,k}H_{\min,k}k}\right)\mathbf{W}_{ij} + \bar{\delta}_{ijV_{\max,k}H_{\min,k}k}\,\mathbf{M}\Bigg) - \mathbf{M} < 0 \tag{4.75}$$

**Table 4.3** IT2 FMB controllers parameters—Part 1

| | |
|---|---|
| Positive definitive matrix | $\mathbf{X} = \begin{bmatrix} 5.5337 & 0.0387 & 0.9308 & -0.0127 \\ 0.0387 & 0.3967 & 0.0277 & 0.0158 \\ 0.9308 & 0.0277 & 1.1754 & -0.0681 \\ -0.0127 & 0.0158 & -0.0681 & 0.0994 \end{bmatrix}$ |

$$\sum_{i=1}^{p}\sum_{j=1}^{c}\Bigg(\bar{\delta}_{ij V_{\max,k} H_{\max,k} k}\,\mathbf{Q}_{ij} - \left(\underline{\delta}_{ij V_{\max,k} H_{\max,k} k} - \bar{\delta}_{ij V_{\max,k} H_{\max,k} k}\right)\mathbf{W}_{ij} + \bar{\delta}_{ij V_{\max,k} H_{\max,k} k}\,\mathbf{M}\Bigg) - \mathbf{M} < 0 \tag{4.76}$$

The IT2 T-S fuzzy controller parameters are generally obtained by solving the LMI stability conditions (4.54)–(4.57). Following the above modification, the LMI conditions become (4.54), (4.55), (4.56), (4.73)–(4.76), and solving them produces the controller parameters. The resulting positive definite matrix $\mathbf{X}$, slack matrices $\mathbf{M}$, $W_{ij}$, and feedback gains $\mathbf{G}_j$ $(i = 1, 2, 3;\ j = 1, 2)$ are presented in Tables 4.3, 4.4 and 4.5.

### *4.5.1 Closed Loop Simulation*

Building an IT2 FMB controller that should force the closed-loop system trajectory to converge at the operational point (trim point) if we start at any random position near it is the stabilization challenge that interests us. If the initial and operating points coincide, the closed-loop system trajectory is likely to stay at this operating point for all subsequent times [3].

As per new desired flight condition $(V^{desired}, H^{desired})$ state and input variables $(\alpha^{equ}, q^{equ}, \theta^{equ},)$ and $(\delta_c^{equ}, \delta_{sf}^{equ}, \delta_{stf}^{equ})$ to the trimmed equilibrium are determined. Rewriting state and input parameters as

$$\mathbf{x}^{\text{trim}} = \left[V^{\text{trim}}\ \alpha^{\text{trim}}\ q^{\text{trim}}\ \theta^{\text{trim}}\right]^{\top} \tag{4.77}$$

$$= \left[V - V^{\text{demand}}\ \alpha - \alpha^{\text{equ}}\ q - q^{\text{equ}}\ \theta - \theta^{\text{equ}}\right]^{\top} \tag{4.78}$$

$$\begin{aligned}\mathbf{u}^{\text{trim}} &= \left[\delta_c^{\text{trim}}\ \delta_{sf}^{\text{trim}}\ \delta_{stf}^{\text{trim}}\right]^{\top} \\ &= \left[\delta_c - \delta_c^{\text{equ}}\ \delta_{sf} - \delta_{sf}^{\text{equ}}\ \delta_{stf} - \delta_{stf}^{\text{equ}}\right]^{\top}\end{aligned} \tag{4.79}$$

**Table 4.4** IT2 FMB controllers parameters—Part 2

| | |
|---|---|
| Slack matrices | $\mathbf{M}=\begin{bmatrix}0.3201 & 0.0097 & -0.0003 & -0.0069\\ 0.0097 & 0.3130 & -0.0086 & 0.0018\\ -0.0003 & -0.0086 & 0.3170 & 0.0102\\ -0.0069 & 0.0018 & 0.0102 & 0.3731\end{bmatrix}$ |
| | $\mathbf{W}_{11}=\begin{bmatrix}5.2515 & 0.8904 & -0.5319 & 0.5548\\ 0.8904 & 4.9683 & -0.2606 & 0.1131\\ -0.5319 & -0.2606 & 4.7664 & -0.7686\\ 0.5548 & 0.1131 & -0.7686 & 4.9410\end{bmatrix}$ |
| | $\mathbf{W}_{12}=\begin{bmatrix}4.9196 & 0.0388 & -0.0231 & 0.0383\\ 0.0388 & 4.8858 & -0.0842 & 0.0389\\ -0.0231 & -0.0842 & 4.8927 & -0.1510\\ 0.0383 & 0.0389 & -0.1510 & 4.7289\end{bmatrix}$ |
| | $\mathbf{W}_{21}=\begin{bmatrix}4.2445 & -0.4450 & 0.0155 & 0.0346\\ -0.4450 & 5.0934 & 0.9888 & -0.3411\\ 0.0155 & 0.9888 & 4.6917 & -0.9210\\ 0.0346 & -0.3411 & -0.9210 & 4.4911\end{bmatrix}$ |
| | $\mathbf{W}_{22}=\begin{bmatrix}4.8320 & -1.5765 & -0.1844 & -0.6972\\ -1.5765 & 5.6809 & 1.2288 & 0.5303\\ -0.1844 & 1.2288 & 5.5090 & 0.5121\\ -0.6972 & 0.5303 & -0.5121 & 4.5963\end{bmatrix}$ |
| | $\mathbf{W}_{31}=\begin{bmatrix}5.5790 & -0.2981 & 0.0047 & 0.1662\\ -0.2981 & 5.1435 & -0.1986 & -0.2026\\ 0.0047 & -0.1986 & 5.5620 & 0.0927\\ 0.1662 & -0.2026 & 0.0927 & 1.8157\end{bmatrix}$ |
| | $\mathbf{W}_{32}=\begin{bmatrix}5.3747 & 0.1989 & 0.1346 & 1.5958\\ 0.1989 & 5.5791 & -0.0131 & 0.0326\\ 0.1346 & -0.0131 & 4.8214 & 0.7670\\ 1.5958 & 0.0326 & 0.7670 & 5.3539\end{bmatrix}$ |

**Table 4.5** IT2 FMB controllers parameters—Part 3

| | |
|---|---|
| Feedback gains | $\mathbf{G}_1=\begin{bmatrix}3.8944 & -41.51 & -20.92 & -200.99\\ 1.3174 & 0.3739 & -9.1067 & -117.99\\ 22.23 & 341.75 & -90.39 & -663.49\end{bmatrix}$ |
| | $\mathbf{G}_2=\begin{bmatrix}-1.4120 & -61.86 & 6.557 & -31.02\\ -0.376 & -31.68 & -1.347 & -187.5\\ -8.1168 & 310.22 & 53.62 & 767.14\end{bmatrix}$ |

**Table 4.6** Trimmed equilibrium flight conditions

| Flight condition | Aircraft velocity (ft/s) | Angle of attack (deg) | Pitch rate (deg/s) | Pitch angle (deg) |
|---|---|---|---|---|
| 1 | 284 | 8.60 | 0.258 | 8.60 |
| 2 | 726 | 4.10 | 0.70 | 4.10 |
| 3 | 977 | 2.80 | 0.90 | 2.80 |

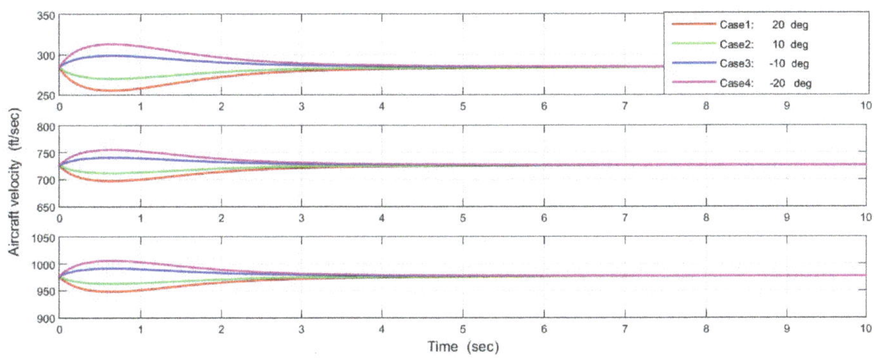


**Fig. 4.3** Control results of aircraft velocity at trimmed speeds of 284, 726, and 977 ft/s for 4 initial conditions

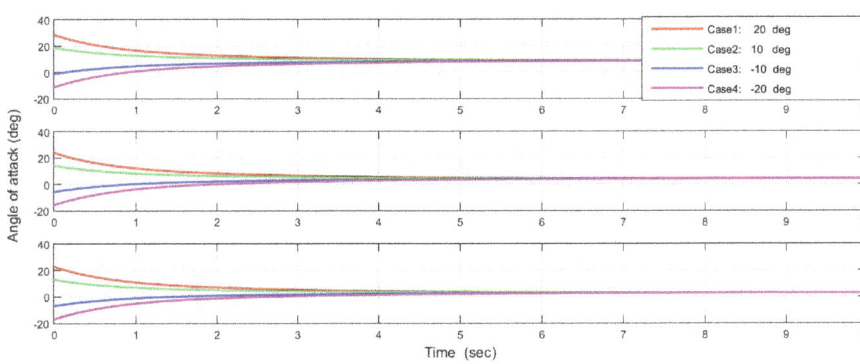


**Fig. 4.4** Control results of Angle of attack at trimmed values 8.6, 4.1, and 2.8 deg for 4 initial conditions

Now for the regulation problem, $\mathbf{x} \rightarrow 0$ represents:

$$V \rightarrow V^{\text{demand}}, \quad \alpha \rightarrow \alpha^{\text{equ}}, \quad q \rightarrow q^{\text{equ}}, \quad \theta \rightarrow \theta^{\text{equ}}$$

Simulations of the closed-loop state-feedback IT2 fuzzy controller are performed under the trim equilibrium flight conditions provided in Table 4.6.

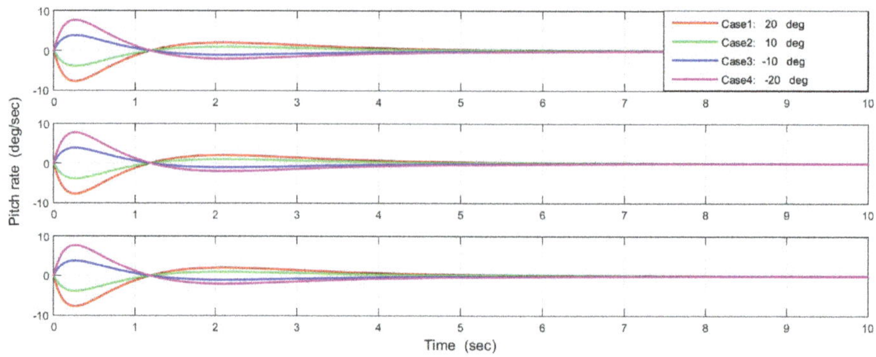


**Fig. 4.5** Control results of pitch rate at trimmed value 0.258, 0.70, and 0.90 deg/s for 4 initial conditions

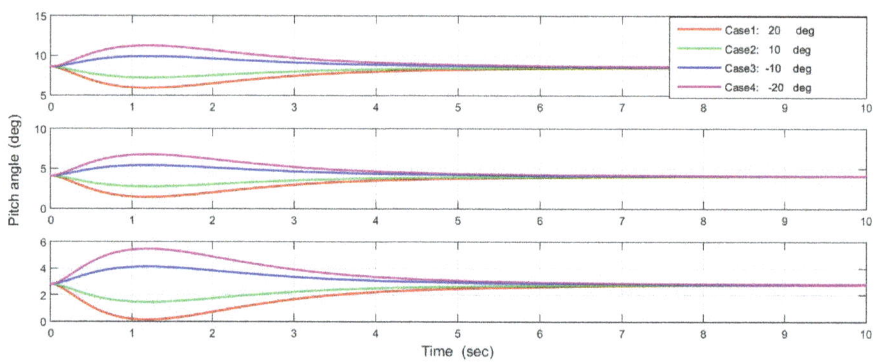


**Fig. 4.6** Control results of Pitch angle at trimmed values 8.6, 4.1, and 2.8 deg for 4 initial conditions

Figures 4.3, 4.4, 4.5 and 4.6 depict the closed-loop behavior of the IT2 T-S FMB controller under different initial angles of attack, $\alpha_{\text{int}}$. Four specific cases of $\alpha_{\text{int}}$ were chosen for the simulations:

$$\begin{aligned}
\text{Case 1:} \quad & \alpha_{c_{\text{int}}} = 20\,\text{deg} \\
\text{Case 2:} \quad & \alpha_{c_{\text{int}}} = 10\,\text{deg} \\
\text{Case 3:} \quad & \alpha_{c_{\text{int}}} = -10\,\text{deg} \\
\text{Case 4:} \quad & \alpha_{c_{\text{int}}} = -20\,\text{deg}
\end{aligned}$$

From Figs. 4.3, 4.4, 4.5 and 4.6 depicts that the IT2 FMB controller stabilized the X-29A aircraft to its trimmed equilibrium points within 07 s.

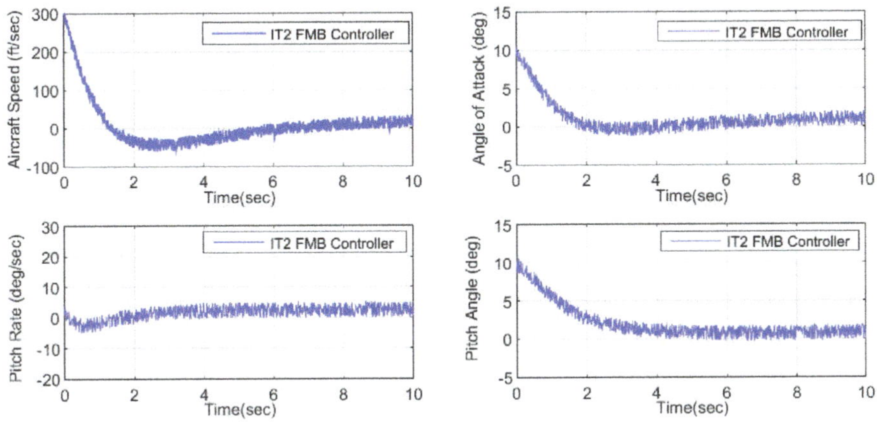


**Fig. 4.7** Condition 1: IT2 FMB controller Robustness Performance

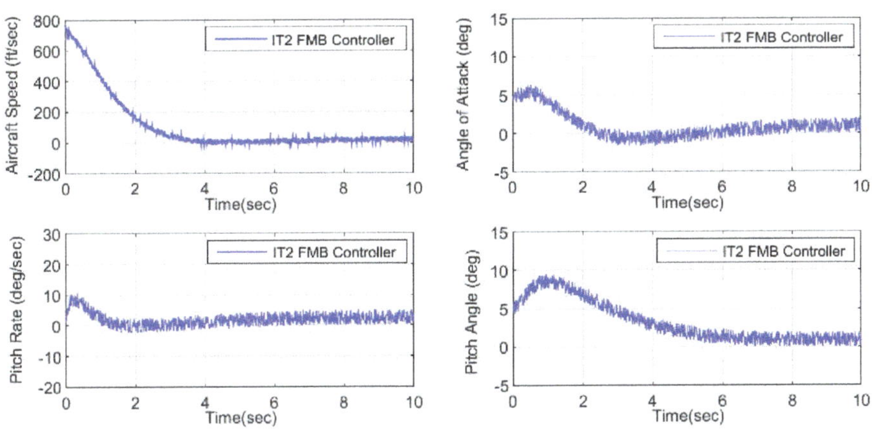


**Fig. 4.8** Condition 2: IT2 FMB controller Robustness Performance

### *4.5.2 Robustness Simulation Results*

The IT2 T-S FMB controller is tested for robustness against disturbances and noise using feedback gain matrices and trim equilibrium flight conditions. White noise with power $10^{-3}$ is added as measurement noise, and simulations are run with a 1 ms sampling time over 10 s. Results (Figs. 4.7, 4.8 and 4.9) show that the controller effectively regulates all four states

$$x = \left[V \ \alpha \ q \ \theta\right]^T$$

with only minor oscillations, which can be suppressed using suitable filters, confirming the controller's robustness.

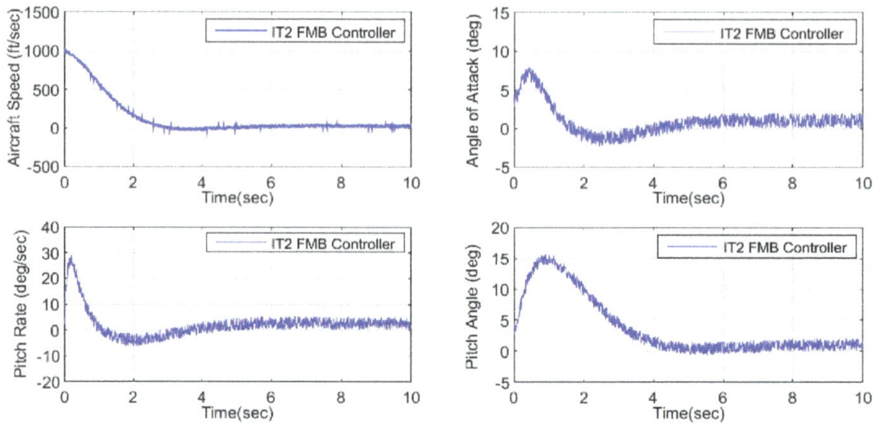


**Fig. 4.9** Condition 3: IT2 FMB controller Robustness Performance

## 4.6 Summary

We have provided a methodical approach to the design of an IT2 T-S FMB control system for general aircraft in this chapter. A new method for mapping the IT2 T-S fuzzy model's premise parameters to the flight envelope parameters (aircraft velocity, operating altitude) is put forth. In order to achieve maximum design flexibility for stability analysis, an imperfect premise matching scenario using the MFD technique is taken into consideration. The nonlinearity resulting from changes in flight conditions is taken into consideration by combining flight envelope features with MFD stability criteria. The IT2 FMB control system is built, and the longitudinal dynamics of the X-29A research aircraft are chosen for modeling and demonstration. The LMI stability requirements are solved to obtain feedback gains and slack matrices. Four initial angle of attack flying scenarios are used to demonstrate the closed-loop response of the suggested IT2 FMB controller at trim equilibrium points. Additionally, fuzzy LQR and LQR-based optimum controllers are compared to the developed IT2 FMB controller's closed-loop performance. The simulation results demonstrate the developed IT2 FMB controller's utility for both aircraft and UAV applications. The real-time implementation of IT2 FMB controller design methods for medium-class UAVs is a suggested future project.

## References


1. M. Biglarbegian, W. W. Melek and J. M. Mendel, "On the Stability of Interval Type-2 TSK Fuzzy Logic Control Systems," in IEEE Transactions on Systems, Man, and Cybernetics, Part B (Cybernetics), vol. 40, no. 3, pp. 798-818, 2010, https://doi.org/10.1109/TSMCB.2009.2029986.

2. B.L. Stevens and F.L. Lewis, "Aircraft Control and Simulation," John Wiley & Sons, 2003.
3. M.Tanaka, K. Yamaguchi, D. Ogura, Y. J. Chen and K. Tanaka. "Non-linear Control of F16 Aircraft via Multiple Nonlinear Model Generation for Any Trimmed Equilibriums." in International Journal of Fuzzy Sys-tems, Vol 16, No. 2, June 2014.
4. D. J. Singh, P. Agrawal, N. K. Verma, A. K. Ghosh, and A. Malagauda-navar, "Interval type-2 TS fuzzy model for angle of attack sensor of the Aircraft", J. Intell. Fuzzy Syst., vol. 34, no. 6, pp. 3891–3901, 2018.
5. X. Yang, H.-K. Lam, and L. Wu, "Membership-dependent stability conditions for type-1 and interval type-2 T-S fuzzy systems," Fuzzy Sets Syst., vol. 356, pp. 44-62, 2019. https://doi.org/10.1016/j.fss.2018.01.018.
6. E. Kayacan and R. Maslim, "Type-2 Fuzzy Logic Trajectory Tracking Control of Quadrotor VTOL Aircraft With Elliptic Membership Functions," in IEEE/ASME Transactions on Mechatronics, vol. 22, no. 1, pp. 339-348, 2017, https://doi.org/10.1109/TMECH.2016.2614672.
7. A. Sarabakha, C. Fu, E. Kayacan and T. Kumbasar, "Type-2 Fuzzy Logic Controllers Made Even Simpler: From Design to Deployment for UAVs," in IEEE Transactions on Industrial Electronics, vol. 65, no. 6, pp. 5069-5077, 2018, https://doi.org/10.1109/TIE.2017.2767546.
8. H. K. Lam and L. D. Seneviratne, "Stability Analysis of Interval Type-2 Fuzzy-Model-Based Control Systems," in IEEE Transactions on Systems, Man, and Cybernetics, Part B (Cybernetics), vol. 38, no. 3, pp. 617-628, 2008, https://doi.org/10.1109/TSMCB.2008.915530.
9. A. Amirkhani, M. Shirzadeh and T. Kumbasar, "Interval Type-2 Fuzzy Cognitive Map-Based Flight Control System for Quadcopters," Int. J. Fuzzy Syst, 2020, https://doi.org/10.1007/s40815-020-00940-8.
10. H. K. Lam, A review on stability analysis of continuous-time fuzzy-model-based control systems: From membership-function-independent to membership-function-dependent analysis. Engineering Applications of Artificial Intelligence, vol 68, pp. 390-408, 2018, https://doi.org/10.1016/j.engappai.2017.09.007.
11. H. K. Lam and M. Narimani, "Stability Analysis and Performance De-sign for Fuzzy-Model-Based Control System Under Imperfect Premise Matching," in IEEE Transactions on Fuzzy Systems, vol. 17, no. 4, pp. 949-961, 2009, https://doi.org/10.1109/TFUZZ.2008.928600.
12. H. K. Lam, L. Wu and Y. Zhao, "Linear matrix inequalities-based membership function-dependent stability analysis for non-parallel distributed compensation fuzzy-model-based control systems," in IET Control Theory & Applications, vol. 8, no. 8, pp. 614-625, 15 May 2014, https://doi.org/10.1049/iet-cta.2013.0171.
13. H. K. Lam, B. Xiao, Y. Yu, X. Yin, H. Han, S.H. Tsai, and C.S. Chen, "Membership-Function-Dependent Stability Analysis and Control Synthesis of Guaranteed Cost Fuzzy-Model-Based Control Systems," International Journal of Fuzzy Systems, vol. 18, 2016, https://doi.org/10.1007/s40815-016-0162-4.
14. H. K. Lam, H. Li, C. Deters, E. L. Secco, H. A. Wurdemann, and K. Althoefer, "Control Design for Interval Type-2 Fuzzy Systems Under Imperfect Premise Matching," in IEEE Transactions on Industrial Electronics, vol. 61, no. 2, pp. 956-968, 2014, https://doi.org/10.1109/TIE.2013.2253064.
15. J.T. Bosworth, "Linearized Aerodynamic and Control Law Models of the X-29A Airplane and Comparison With Flight Data." NASA TM-4356, 1992.
16. H. K. Lam and M. Narimani, "Quadratic-Stability Analysis of Fuzzy-Model-Based Control Systems Using Staircase Membership Functions," in IEEE Transactions on Fuzzy Systems, vol. 18, no. 1, pp. 125-137, 2010, https://doi.org/10.1109/TFUZZ.2009.2037744.
17. H. K. Lam, "Polynomial Fuzzy-Model-Based Control Systems: Stability Analysis Via Piecewise-Linear Membership Functions," in IEEE Transactions on Fuzzy Systems, vol. 19, no. 3, pp. 588-593, 2011, https://doi.org/10.1109/TFUZZ.2011.2118215.
18. O.A. Arqub and M. Al-Smadi, "Fuzzy conformable fractional differential equations: novel extended approach and new numerical solutions," Soft Comput, vol. 24, pp. 12501–12522, 2020, https://doi.org/10.1007/s00500-020-04687-0.
19. O.A. Arqub, "Adaptation of reproducing kernel algorithm for solving fuzzy Fredholm–Volterra integrodifferential equations," Neural Comput & Applic, vol. 28, pp. 1591–1610, 2017, https://doi.org/10.1007/s00521-015-2110-x.

20. O.A. Arqub, M. Al-Smadi and S. Momani et al., "Numerical solutions of fuzzy differential equations using reproducing kernel Hilbert space method," Soft Comput, vol. 20, pp. 3283–3302, 2016, https://doi.org/10.1007/s00500-015-1707-4.
21. O.A. Arqub, M. Al-Smadi and S. Momani et al., "Application of reproducing kernel algorithm for solving second-order, two-point fuzzy boundary value problems," Soft Comput, vol. 21, pp. 7191–7206, 2017, https://doi.org/10.1007/s00500-016-2262-3.
22. R. J. Caverly and J. Forbes "LMI Properties and Applications in Systems, Stability, and Control Theory." ArXiv abs/1903.08599v3 (Apr 2021): n. pag.
23. F. D. Bianchi, H. De. Battista and R. J. Mantz, "Wind Turbine Control Systems Principles, Modelling and Gain Scheduling Design", Springer-Verlag London 2007, ISBN 978-1-84628-493-9, https://doi.org/10.1007/1-84628-493-7.

# Chapter 5
# Robustness of General Type-2 T-S Fuzzy System

**Abstract** This chapter presents the robustness of General Type-2 Fuzzy Logic System (GT2 FLS) as an extension of Interval Type-2 (IT2) FLS. The theoretical base and mathematical derivations developed for the robustness of IT2 FLS are extended for GT2 FLS using $\alpha$-plane representation. The robustness analysis for computing the robustness index is modified for $\alpha$ level type-2 FLS, and by aggregating all such levels, it is further customized for GT2 FLS. To demonstrate the efficacy of extended methodology, function approximation-based simulation examples are considered. It was understood that, similar to T1 and IT2 FLS, the GT2 FLS can be developed to attain robust performance with a more flexible structure. The additional design flexibility associated with GT2 FLS provides reduced output error and enhanced robustness performance based on the nature of the application.

## 5.1 Introduction

In the fuzzy systems design, researchers have recently been interested in devising the optimum tolerance of the system when it is exposed to minor deviations of the specimen points. This analysis becomes a prerequisite for examining the effect of input variations on the anticipated output. The above investigation is referred to as robustness analysis. In literature, various theoretical developments have emerged to examine the robustness features with respect to fuzzy operators [1, 2], parameterized formulation of fuzzy inference mechanism [3], by using statistics and probabilistic approach, and algebraic operators [4–8]. Though these approaches have initiated the robustness of FLS, they lack a systematic methodology and are limited to fuzzy operators. Further to overcome the above limitations and in line with practical applicability, Biglarbegian et al. have presented a systematic approach for robustness analysis. In [9, 10], the robustness of IT2 FLS is thoroughly investigated, and a metric for measuring the robustness of FLS is demonstrated for various case studies related to fuzzy modeling. The usual structure of IT2 TSK FLSs is utilized for robustness investigation, and it was found that the developed approach is well-suited and applicable for T1 and IT2 FLSs design applications.

 
D. J. Singh et al., *Fuzzy Systems with Applications to Modeling and Control of an Aircraft*, Algorithms for Intelligent Systems, https://doi.org/10.1007/978-981-95-7888-7_5

Inspired by their proposed work and mathematical foundation, the authors try to extend the theoretical base of [9, 10] for General Type-2 FLSs. The mathematical derivations presented for IT2 FLS are utilized as a baseline, and further formulation for robustness of GT2 FLSs is explored. The $\alpha$-plane representation of GT2 FLS is adopted for robustness analysis [11, 12]. The extended algorithms for robustness analysis of GT2 FLS are applied on similar case studies as presented in [10]. The simulation results reveal that GT2 FLS exhibit enhanced robust behavior. The computational complexity associated with robustness analysis can be suitably simplified by the optimum choice of the number of $\alpha$-planes.

The manuscript is organized as follows. Section 5.2 provides an outline about GT2 FLS with $\alpha$-plane representation. Section 5.3 gives the mathematical description for robustness analysis of GT2 FLS based on the theoretical background established for IT2 FLS. Section 5.4 demonstrates the simulation studies. Finally, Sect. 5.5 concludes the manuscript.

## 5.2 General Type-2 Fuzzy Sets with Alpha Plane Representation

An IT2 TSK FLS is described by IF-THEN fuzzy rules that signify input-output associations in the given arrangement ($i$th-rule) as follows

$$\begin{aligned} R^i:\ & \text{IF } x_1 \text{ is } \tilde{F}_1^i \text{ and } x_2 \text{ is } \tilde{F}_2^i \text{ and } \ldots \text{ and } x_p \text{ is } \tilde{F}_p^i \\ & \text{THEN } y^i(x) = C_1^i x_1 + C_2^i x_2 + \ldots + C_p^i x_p + C_0^i. \end{aligned} \tag{5.1}$$

The IT2 TSK FLS parameters $C_j^i$ and $C_0^i$ represent the consequent part with an interval set. These parameters are given as [13] $C_j^i = [v_j^i - s_j^i, v_j^i + s_j^i]$ and $C_0^i = [v_0^i - s_0^i, v_0^i + s_0^i]$. Here, $v_j^i$ indicates the center (mean) of $C_j^i$ and $s_j^i$ indicates the spread of $C_j^i$. The number of rules is $i = 1, \ldots., M$ and $\tilde{F}_j^i$ represents the IT2 FS of input state $j$ in rule $i$ with $j = 1, \ldots, P$. Thus, rule consequent $R^i$ is stated as $Y^i = [y_l^i, y_r^i]$ where,

$$y_l^i = \sum_{k=1}^{p} \mathbf{v}_j^i \mathbf{x}_k + \mathbf{v}_0^i - \sum_{k=1}^{p} s_j^i |\mathbf{x}_k| - s_0^i \tag{5.2}$$

and

$$y_r^i = \sum_{k=1}^{p} \mathbf{v}_j^i \mathbf{x}_k + v_0^i + \sum_{k=1}^{p} s_j^i |\mathbf{x}_k| + s_0^i. \tag{5.3}$$

The above formulation of IT2 TSK FLS is further employed for the development of GT2 TSK FLS. The Secondary Membership Function (SMF) of GT2 FLS is triangular in nature and it can be characterized by its $\alpha$- plane decomposition. The base of SMF equals $\bar{\mu}_{\tilde{F}}(x) - \underline{\mu}_{\tilde{F}}(x)$ and its apex location is attained as [11],

$$\text{apex}\,(x) = \underline{\mu}_{\tilde{F}}(x) + \omega\left[\overline{\mu}_{\tilde{F}}(x) - \underline{\mu}_{\tilde{F}}(x)\right] \tag{5.4}$$

where $\omega \in [0, 1]$ is the parameter of design and it describes the nature of SMF. The $\alpha$-cut of SMF are as per [11]

$$\begin{aligned}\tilde{F}_\alpha(x) &= [a_\alpha(x), b_\alpha(x)]\\ a_\alpha(x) &= \underline{\mu}_{\tilde{F}}(x) + \omega[\overline{\mu}_{\tilde{F}}(x) - \underline{\mu}_{\tilde{F}}(x)]\alpha \\ b_\alpha(x) &= \overline{\mu}_{\tilde{F}}(x) - (1-\omega)[\overline{\mu}_{\tilde{F}}(x) - \underline{\mu}_{\tilde{F}}(x)]\alpha\end{aligned} \quad . \tag{5.5}$$

The $\alpha$-cut decomposition (for each rule) $\tilde{F}_j^i(x_j')$ is derived as

$$\tilde{F}_j^i(x_j') = \sup_{\forall \alpha \in [0,1]} \alpha \Big/ \left[a_{j,\alpha}^i(x_j'), b_{j,\alpha}^i(x_j')\right]. \tag{5.6}$$

The firing interval $\tilde{F}_\alpha^i(x')$ for level $\alpha$ is,

$$\begin{aligned}\tilde{F}_\alpha^i(\mathbf{x}') &= \left[\underline{f}_\alpha^{i'}(\mathbf{x}'), \overline{f}_\alpha^i(\mathbf{x}')\right]\\ \underline{f}_\alpha^{i'}(\mathbf{x}') &= T_{j=1}^p \mathbf{a}_{j,\alpha}^i(x_j')\\ \overline{f}_\alpha^i(\mathbf{x}') &= T_{j=1}^p \mathbf{b}_{j,\alpha}^i(x_j')\end{aligned} \tag{5.7}$$

Discretize the $\alpha$- cuts for $\lambda = 1, \ldots.., \rho$ stages. Then by aggregating the all-horizontal slices $\alpha_1, \alpha_2, \ldots., \alpha_\lambda$ SMF are created. The steps for designing of GT2 FLS are described as:

1. For all the vertical slices of $\tilde{F}$ , chose the $\alpha$- cuts number $\rho$. Then discretize the $\alpha_\lambda$ values as $\alpha_1, \alpha_2, \ldots., \alpha_\rho$ for $\lambda = 1, \ldots., \rho$ levels.
2. Over the support of $\tilde{F}$, compute

$$\left[a_{\alpha_\lambda}^i(\mathbf{x}^*),\; b_{\alpha_\lambda}^i(\mathbf{x}^*)\right] \quad \text{for } \lambda = 1, \ldots, \rho \quad \text{as:}$$

$$\begin{aligned}a_{\alpha_\lambda}^i(\mathbf{x}^*) &= \underline{\mu}_{\tilde{F}_i}(\mathbf{x}^*) + \omega[\overline{\mu}_{\tilde{F}_i}(\mathbf{x}^*) - \mu_{\tilde{F}_i}(\mathbf{x}^*)]\alpha_\lambda\\ b_{\alpha_\lambda}^i(\mathbf{x}^*) &= \underline{\mu}_{\tilde{F}_i}(\mathbf{x}^*) - (1-\omega)[\overline{\mu}_{\tilde{F}_i}(\mathbf{x}^*) - \underline{\mu}_{\tilde{F}_i}(\mathbf{x}^*)]\alpha_\lambda.\end{aligned} \tag{5.8}$$

3. Find the $\tilde{F}_{\alpha_\lambda}^i(\mathbf{x}^*)$ (as a firing interval) for each level $\alpha_\lambda$ corresponding to $i$th rule as

$$\tilde{F}^i_{\alpha_\lambda}(\mathbf{x}^*) \equiv \left[\underline{f}^i_{\alpha_\lambda}(\mathbf{x}^*), \overline{f}^i_{\alpha_\lambda}(\mathbf{x}^*)\right] \text{ where, } \left[\underline{f}^i_{\alpha_\lambda}(\mathbf{x}^*), \overline{f}^i_{\alpha_\lambda}(\mathbf{x}^*)\right] \equiv \left[a^i_{\alpha_\lambda}(\mathbf{x}^*), b^i_{\alpha_\lambda}(\mathbf{x}^*)\right]. \tag{5.9}$$

4. Calculate the $m-n$ outputs $y_{mn,\alpha_\lambda}(\mathbf{x}^*)$ at every $\alpha_1, \alpha_2, \ldots, \alpha_\rho$ as [11],

$$y_{mn,\alpha_\lambda}(\mathbf{x}^*) = m\frac{\sum_{i=1}^{M} y_l^i \underline{f}^i_{\alpha_\lambda}(\mathbf{x}^*)}{\sum_{i=1}^{M} \underline{f}^i_{\alpha_\lambda}(\mathbf{x}^*)} + n\frac{\sum_{i=1}^{M} y_r^i \overline{f}^i_{\alpha_\lambda}(\mathbf{x}^*)}{\sum_{i=1}^{M} \overline{f}^i_{\alpha_\lambda}(\mathbf{x}^*)}. \tag{5.10}$$

   Here, rule consequent $R^i$, i.e. $\left[y_l^i, y_r^i\right]$ are estimated from [2, 3].
5. By defuzzifying the $y_{mn,\alpha_\lambda}(\mathbf{x}^*)$ for each $\alpha_1, \alpha_2, \ldots, \alpha_\rho$ levels the GT2 FLS are obtained as

$$y_{mnGT2}(\mathbf{x}^*) = \frac{\sum_{\lambda=1}^{\rho} \alpha_\lambda y_{mn,\alpha_\lambda}(\mathbf{x}^*)}{\sum_{\lambda=1}^{\rho} \alpha_\lambda}. \tag{5.11}$$

6. The values of $m, n$ parameters are optimized by using gradient descent algorithms,

$$\text{MSE} = \frac{1}{N}\sum_{k=1}^{N}\left(y_{mn,GT2}(\mathbf{x}_k) - y_k\right)^2. \tag{5.12}$$

## 5.3 Robustness of General Type-2 FLS

In this section, the robustness governing equations derived for IT2 FLS are extended for GT2 FLS. The $\alpha_\lambda$ level $m-n$ outputs $y_{mn,\alpha_\lambda}(\mathbf{x}^*)$ for GT2 FLS is given as

$$y_{m,\alpha_j}\left(\mathbf{x}^*\right) \equiv m\frac{\sum_{i=1}^{M} y_l^i \underline{f}^i_{\alpha_\lambda}(\mathbf{x}^*)}{\sum_{i=1}^{M} \underline{f}^i_{\alpha_\lambda}(\mathbf{x}^*)} + n\frac{\sum_{i=1}^{M} y_r^i \bar{f}^i_{\alpha_\lambda}(\mathbf{x}^*)}{\sum_{i=1}^{M} \bar{f}^i_{\alpha_\lambda}(\mathbf{x}^*)}. \tag{5.13}$$

The resultant output of GT2 FLS is obtained by aggregating the each $\alpha_1$, $\alpha_2$, ....., $\alpha_\rho$ level outputs as

$$Y(\mathbf{x}^*) = \sum_{\lambda=1}^{\rho} y_{m,\alpha_\lambda}(\mathbf{x}^*) \tag{5.14}$$

$$Y(\mathbf{x}^*) = \sum_{\lambda=1}^{\rho}\left[m\frac{\sum_{i=1}^{M} y_l^i \underline{f}^i_{\alpha_\lambda}(\mathbf{x}^*)}{\sum_{i=1}^{M} \underline{f}^i_{\alpha_\lambda}(\mathbf{x}^*)} + n\frac{\sum_{i=1}^{M} y_r^i \bar{f}^i_{\alpha_\lambda}(\mathbf{x}^*)}{\sum_{i=1}^{M} \bar{f}^i_{\alpha_\lambda}(\mathbf{x}^*)}\right] \tag{5.15}$$

Now, define the $\alpha_\lambda$ level normalized upper and lower firing strength as $\underline{w}^i_{\alpha_\lambda}(\mathbf{x}^*)$ and $\bar{w}^i_{\alpha_\lambda}(\mathbf{x}^*)$ for the $i$th rule.

Where,

$$\underline{w}^i_{\alpha_\lambda}(\mathbf{x}^*) = \frac{\underline{f}^i_{\alpha_\lambda}(\mathbf{x}^*)}{\sum_{j=1}^{M} \underline{f}^j_{\alpha_\lambda}(\mathbf{x}^*)}$$

$$\bar{w}^i_{\alpha_\lambda}(\mathbf{x}^*) = \frac{\bar{f}^i_{\alpha_\lambda}(\mathbf{x}^*)}{\sum_{j=1}^{M} \bar{f}^j_{\alpha_\lambda}(\mathbf{x}^*)}$$

Now Eq. (5.15) are further elaborated as

$$Y(\mathbf{x}^*) = m\sum_{\lambda=1}^{\rho}\sum_{i=1}^{M} \underline{w}^i_{\alpha_\lambda}(\mathbf{x}^*)y^i_l(\mathbf{x}^*) + n\sum_{\lambda=1}^{\rho}\sum_{i=1}^{M} \bar{w}^i_{\alpha_\lambda}(\mathbf{x}^*)y^i_r(\mathbf{x}^*) \tag{5.16}$$

The output (deviated) $\Delta Y$ is estimated by assuming the small deviation in the input $\mathbf{x}^*$, as follows:

$$\begin{aligned}
\Delta Y &= Y(\mathbf{x}^* + \Delta\mathbf{x}^*) - Y(\mathbf{x}^*)\\
\Delta Y &= m\sum_{\lambda=1}^{\rho}\sum_{i=1}^{M} \underline{w}^i_{\alpha_\lambda}(\mathbf{x}^* + \Delta\mathbf{x}^*)y^i_l(\mathbf{x}^* + \Delta\mathbf{x}^*)\ldots.\\
&\ldots + n\sum_{\lambda=1}^{\rho}\sum_{i=1}^{M} \bar{w}^i_{\alpha_\lambda}(\mathbf{x}^* + \Delta\mathbf{x}^*)y^i_r(\mathbf{x}^* + \Delta\mathbf{x}^*)\ldots.\\
&\ldots.. - m\sum_{\lambda=1}^{\rho}\sum_{i=1}^{M} \underline{w}^i_{\alpha_\lambda}(\mathbf{x}^*)y^i_l(\mathbf{x}^*) - n\sum_{\lambda=1}^{\rho}\sum_{i=1}^{M} \bar{w}^i_{\alpha_\lambda}(\mathbf{x}^*)y^i_r(\mathbf{x}^*)
\end{aligned} \tag{5.17}$$

Define the notation $\Delta A$ and $\Delta B$ as

$$\begin{aligned}
\Delta A &= m\sum_{\lambda=1}^{\rho}\sum_{i=1}^{M} \underline{w}^i_{\alpha_\lambda}(\mathbf{x}^* + \Delta\mathbf{x}^*)y^i_l(\mathbf{x}^* + \Delta\mathbf{x}^*) - m\sum_{\lambda=1}^{\rho}\sum_{i=1}^{M} \underline{w}^i_{\alpha_\lambda}(\mathbf{x}^*)y^i_l(\mathbf{x}^*)\\
\Delta B &= n\sum_{\lambda=1}^{\rho}\sum_{i=1}^{M} \bar{w}^i_{\alpha_\lambda}(\mathbf{x}^* + \Delta\mathbf{x}^*)y^i_r(\mathbf{x}^* + \Delta\mathbf{x}^*) - n\sum_{\lambda=1}^{\rho}\sum_{i=1}^{M} \bar{w}^i_{\alpha_\lambda}(\mathbf{x}^*)y^i_r(\mathbf{x}^*)
\end{aligned} \tag{5.18}$$

With these new notations, the output $\Delta Y$ is expressed as follows:

$$\Delta Y = \Delta A + \Delta B \tag{5.19}$$

By assuming small $\Delta x$ the component of equation $\Delta A$ and $\Delta B$ can be further simplified as follows:

$$
\begin{aligned}
\Delta A &= m \sum_{\lambda=1}^{\rho} \sum_{i=1}^{M} \left[ \underline{w}^i_{\alpha_\lambda} \sum_{j=1}^{P} \frac{\partial y^i_l}{\partial x_j} \Delta x_j + y^i_l \sum_{j=1}^{P} \frac{\partial \underline{w}^i_{\alpha_\lambda}}{\partial x_j} \Delta x_j \right] \\
\Delta B &= n \sum_{\lambda=1}^{\rho} \sum_{i=1}^{M} \left[ \bar{w}^i_{\alpha_\lambda} \sum_{j=1}^{P} \frac{\partial y^i_r}{\partial x_j} \Delta x_j + y^i_r \sum_{j=1}^{P} \frac{\partial \bar{w}^i_{\alpha_\lambda}}{\partial x_j} \Delta x_j \right]
\end{aligned}
\tag{5.20}
$$

The deviated output can be further represented as

$$
\begin{aligned}
\Delta Y = {} & m \sum_{\lambda=1}^{\rho} \sum_{i=1}^{M} \left[ \underline{w}^i_{\alpha_\lambda} \sum_{j=1}^{P} \frac{\partial y^i_l}{\partial x_j} \Delta x_j + y^i_l \sum_{j=1}^{P} \frac{\partial \underline{w}^i_{\alpha_\lambda}}{\partial x_j} \Delta x_j \right] \\
& + n \sum_{\lambda=1}^{\rho} \sum_{i=1}^{M} \left[ \bar{w}^i_{\alpha_\lambda} \sum_{j=1}^{P} \frac{\partial y^i_l}{\partial x_j} \Delta x_j + y^i_l \sum_{j=1}^{P} \frac{\partial \bar{w}^i_{\alpha_\lambda}}{\partial x_j} \Delta x_j \right]
\end{aligned}
\tag{5.21}
$$

The Eq. (5.21) is further simplified as

$$
\begin{aligned}
\Delta Y = {} & \sum_{\lambda=1}^{\rho} \sum_{i=1}^{M} \sum_{j=1}^{P} \left[ m \underline{w}^i_{\alpha_\lambda} \frac{\partial y^i_l}{\partial x_j} \Delta x_j + m y^i_l \frac{\partial \underline{w}^i_{\alpha_\lambda}}{\partial x_j} \Delta x_j \right] \\
& + \sum_{\lambda=1}^{\rho} \sum_{i=1}^{M} \sum_{j=1}^{P} \left[ m \bar{w}^i_{\alpha_\lambda} \frac{\partial y^i_r}{\partial x_j} \Delta x_j + m y^i_r \frac{\partial \bar{w}^i_{\alpha_\lambda}}{\partial x_j} \Delta x_j \right]
\end{aligned}
\tag{5.22}
$$

$$
\Delta Y = \sum_{\lambda=1}^{\rho} \sum_{i=1}^{M} \sum_{j=1}^{P} \left[ m \underline{w}^i_{\alpha_\lambda} \frac{\partial y^i_l}{\partial x_j} + m y^i_l \frac{\partial \underline{w}^i_{\alpha_\lambda}}{\partial x_j} + m \bar{w}^i_{\alpha_\lambda} \frac{\partial y^i_r}{\partial x_j} + m y^i_r \frac{\partial \bar{w}^i_{\alpha_\lambda}}{\partial x_j} \right] \Delta x_j
\tag{5.23}
$$

From the above equation, define the term $X^l_j$ and $X^u_j$ as follows:

$$
\begin{aligned}
X^l_j &\equiv \sum_{\lambda=1}^{\rho} \sum_{i=1}^{M} \left[ m \underline{w}^i_{\alpha_\lambda} \frac{\partial y^i_l}{\partial x_j} + m y^i_l \frac{\partial \underline{w}^i_{\alpha_\lambda}}{\partial x_j} \right] \\
X^u_j &\equiv \sum_{\lambda=1}^{\rho} \sum_{i=1}^{M} \left[ n \bar{w}^i_{\alpha_\lambda} \frac{\partial y^i_r}{\partial x_j} + n y^i_r \frac{\partial \bar{w}^i_{\alpha_\lambda}}{\partial x_j} \right]
\end{aligned}
\tag{5.24}
$$

Now Eq. (5.23) is represented as follows:

$$\Delta Y = \sum_{j=1}^{P} X_j^l \Delta x_j + \sum_{j=1}^{P} X_j^u \Delta x_j$$
$$= \sum_{j=1}^{P} \left(X_j^l + X_j^u\right) \Delta x_j \quad (5.25)$$

### 5.3.1 *Robustness Definition*

In this subsection, the robustness definition proposed in [9, 10] is repeated for comprehensiveness. The proposed definition is further extended for GT2 TSK FLS.

The robustness is a multi-objective optimization problem for maximum allowable $|\Delta x_i||$ with the given desired output deviation $\Delta Y_{\text{desired}} \geq 0$.

The mathematical representations are as follows:

$$\begin{cases} \text{Maximize:} & |\Delta x_1, \ldots, \Delta x_p| \\ \text{Subject to:} & \left|\sum_{j=1}^{P} \left(X_j^l + X_j^u\right) \Delta x_j\right| \leq \Delta Y_{\text{desired}} \\ & -\beta_1 \leq \Delta x_1 \leq \gamma_1 \\ & -\beta_2 \leq \Delta x_2 \leq \gamma_2 \\ & \vdots \\ & -\beta_P \leq \Delta x_P \leq \gamma_P \end{cases} \quad (5.26)$$

According to robustness definition proposed in [9, 10] robustness measure for IT2 TSK FLS is given as

$$R_j = \frac{\max |\Delta x_j|}{\Delta Y_{\text{desired}}} \quad (5.27)$$

For computing the robustness index, $|\Delta x_i|$ are obtained for a given $\Delta Y_{desired}$. For every specific input of the FLS, the maximum acceptable deviation is estimated as [9].

$\max |\Delta x_i| = \min |\Delta^i x_i|, i = 1, \ldots.q \; j = 1, \ldots.P$

Here, $|\Delta^i x_i|$ is maximum permissible $j$th input deviation for $i$th training pair.

The robustness measure proposed for IT2 TSK FLS is further modified for GT2 TSK FLS as follows:

1. Determine the $\Delta Y_{\text{desired}}$ as per design requirements.
2. Solve the multi objective optimization problem to find max $|\Delta x_i|_{\alpha_\lambda}$ at selected $\alpha_\lambda$ level.
3. Calculate the robustness measure for $\alpha_\lambda$ level slice as

$$R_j = \frac{\max |\Delta x_j|_{\alpha_\lambda}}{\Delta Y_{\text{desired}}} \tag{5.28}$$

4. Repeat steps 2 and 3 for every $\alpha_1, \alpha_2, \ldots., \alpha_\rho$ levels.
5. Obtain $(R_j)_{GT2}$ by aggregating the $\alpha_\lambda = \alpha_1, \alpha_2, \ldots., \alpha_\rho$ level $R_{j\alpha_\lambda}{}'s$ as per (5.29).

$$(R_j)_{GT2} = \sum_{\lambda=1}^{\rho} R_{j\alpha_\lambda} \tag{5.29}$$

Now for a single input case, Eq. (5.25) will modified as

$$\Delta Y = (X^l + X^u)\Delta x \tag{5.30}$$

$$\frac{\Delta Y}{\Delta x} = \sum_{\lambda=1}^{\rho}\sum_{i=1}^{M}\left[m\underline{w}^i_{\alpha_\lambda}\frac{\partial y^i_l}{\partial x} + m y^i_l \frac{\partial \underline{w}^i_{\alpha_\lambda}}{\partial x}\right] + \ldots \sum_{\lambda=1}^{\rho}\sum_{i=1}^{M}\left[n\bar{w}^i_{\alpha_\lambda}\frac{\partial y^i_r}{\partial x} + n y^i_r \frac{\partial \bar{w}^i_{\alpha_\lambda}}{\partial x}\right] \tag{5.31}$$

Define the parameter $\Delta$, such that

$$\frac{\Delta x}{\Delta Y} = \frac{1}{\Delta_{GT2}}$$

where,

$$\Delta_{GT2} = \sum_{\lambda=1}^{\rho}\sum_{i=1}^{M}\left[m\underline{w}^i_{\alpha_\lambda}\frac{\partial y^i_l}{\partial x} + m y^i_l \frac{\partial \underline{w}^i_{\alpha_\lambda}}{\partial x}\right] + \ldots \sum_{\lambda=1}^{\rho}\sum_{i=1}^{M}\left[n\bar{w}^i_{\alpha_\lambda}\frac{\partial y^i_r}{\partial x} + n y^i_r \frac{\partial \bar{w}^i_{\alpha_\lambda}}{\partial x}\right] \tag{5.32}$$

The maximum permissible deviation in input $x$ is attained when $|\Delta Y| = |\Delta Y_{\text{desired}}|$ i.e.

$$\frac{|\Delta x|_{\max}}{|\Delta Y_{desired}|} = \frac{1}{|\Delta_{GT2}|} = (R_j)_{GT2} \tag{5.33}$$

Similarly, the methodology presented in [9] for the two and multi-input cases can also be modified for calculating the robustness index.

## 5.4 Simulation Studies

In this section, two simulation examples as presented in [10] are considered to demonstrate the robustness of GT2 FLS. The error Performance Index (EPI) and Robustness Performance Index (RPI) proposed in [9, 10] are taken for the performance study.

The EPI and RPI performance parameters for GT2 FLS are estimated in a similar manner as they are estimated for IT2 FLS in [9]. As comparative parameters of FLS performance, a higher value of EPI and RPI indicates more error modeling capability with robust behavior.

### 5.4.1 *Example 1 (Case of Single Input)*

In this case, GT2 TSK FLS is created to approximate a function given as (5.34).

$$f(x) = x + \frac{1}{x^2} \tag{5.34}$$

The set of 75 input-output pairs is selected randomly in [1, 3] as a sample training data. The parameters of tuning $(m, n)$ are selected as (0.88, 1.1 and parameters of Gaussian MFs are as per Table 5.1 of [10].

To carry out a robustness study, various desired deviated outputs are selected, and outcomes are compared with T1 and IT2 T-S FLS in Table 5.1. The simulation results are obtained by taking $\alpha = [0.2, 0.4, 0.6, 0.8, 1.0]$, which shows that GT2 FLS provides improved robust behavior like IT2 FLS.

Further, the set of generated input-output data pairs is added by Gaussian noise having zero (0) mean and values of standard deviation 0.1 to the original data. The simulations are again performed by using the same set of GT2 TSK FLS parameters.

The robustness characteristics of GT2 FLS in the presence of noisy data are demonstrated in Table 5.2. The result shows that GT2 FLS is more suitable for noisy, uncertain data.

**Table 5.1** GT2 FLS Robustness performance with clean data: Example 1

| $\Delta Y_{desired}$ | T1 [10] | | IT2 [10] | | GT2 | |
|---|---|---|---|---|---|---|
| | e | EPI % | e | EPI % | e | EPI % |
| 0.01 | 0.3617 | – | 0.0940 | 74% | 0.0931 | 74.26% |
| 0.05 | 0.3596 | – | 0.1014 | 71.8% | 0.1009 | 71.94% |
| 0.1 | 0.3587 | – | 0.0942 | 73.74% | 0.0935 | 73.93% |
| 0.15 | 0.3506 | – | 0.0839 | 76.08% | 0.0838 | 76.09% |
| 0.2 | 0.3302 | – | 0.0837 | 74.64% | 0.0829 | 74.89% |

**Table 5.2** GT2 FLS Robustness performance with noisy data: Example 1

| $\Delta Y_{desired}$ | T1 [10] | | IT2 [10] | | GT2 | |
|---|---|---|---|---|---|---|
| | e | EPI % | e | EPI % | e | EPI % |
| 0.01 | 0.3834 | – | 0.0977 | 74.5% | 0.0949 | 75.23% |
| 0.05 | 0.3811 | – | 0.1054 | 72.33% | 0.1029 | 72.99% |
| 0.1 | 0.3802 | – | 0.0979 | 74.23% | 0.0953 | 74.91% |
| 0.15 | 0.3716 | – | 0.0872 | 76.52% | 0.0854 | 77.0% |
| 0.2 | 0.3500 | – | 0.0870 | 75.13% | 0.0845 | 75.84% |

## 5.4.2 *Example 2 (Two Input Case)*

In this example, GT2 TSK FLS is developed to approximate a function of two inputs as follows

$$f(x_1, x_2) = (\sin(x_1)/x_1) - (\sin(x_2)/x_2) \tag{5.35}$$

**Table 5.3** GT2 FLS Robustness performance with clean data: Example 2

| $\Delta Y_{desired} = 0.01$ | | | | | | |
|---|---|---|---|---|---|---|
| | e | EPI % | $R_1$ | $R_2$ | $RPI_1\%$ | $RPI_2\%$ |
| T1[10] | 0.1491 | – | 0.0114 | 0.0100 | – | – |
| IT2[10] | 0.1476 | 1.0231% | 0.0166 | 0.0100 | 45.61 | 0 |
| GT2 | 0.1470 | 1.408% | 0.0168 | 0.0110 | 47.36 | 10 |
| $\Delta Y_{desired} = 0.03$ | | | | | | |
| T1[10] | 0.1491 | – | 0.0034 | 0.0033 | – | – |
| IT2[10] | 0.1449 | 2.81% | 0.0034 | 0.0034 | 0 | 3.03 |
| GT2 | 0.1448 | 2.88% | 0.0034 | 0.0034 | 0 | 3.03 |
| $\Delta Y_{desired} = 0.05$ | | | | | | |
| T1[10] | 0.1499 | – | 0.0066 | 0.9400 | – | – |
| IT2[10] | 0.1450 | 3.26% | 0.0081 | 0.5120 | 44.26 | -45.53 |
| GT2 | 0.1448 | 3.40% | 0.0080 | 0.5121 | 21.21 | -45.52 |
| $\Delta Y_{desired} = 0.07$ | | | | | | |
| T1[10] | 0.1615 | – | 0.0021 | 2.8471 | – | – |
| IT2[10] | 0.1534 | 5.04% | 0.0067 | 2.6029 | > 100 | -9.38 |
| GT2 | 0.1531 | 5.20% | 0.0067 | 2.6028 | > 100 | -8.58 |
| $\Delta Y_{desired} = 0.1$ | | | | | | |
| T1[10] | 0.1763 | – | 1 | 3 | – | – |
| IT2[10] | 0.1688 | 4.28% | 1 | 3 | 0 | 0 |
| GT2 | 0.1688 | 4.28% | 1 | 3 | 0 | 0 |

**Table 5.4** GT2 FLS Robustness performance with noisy data: Example 2

| | e | EPI % | $R_1$ | $R_2$ | $RPI_1\%$ | $RPI_2\%$ |
|---|---|---|---|---|---|---|
| $\Delta Y_{desired} = 0.01$ | | | | | | |
| T1[10] | 0.1580 | – | 0.011628 | 0.0103 | – | – |
| IT2[10] | 0.1535 | 2.87% | 0.016932 | 0.0103 | 46.52 | 0 |
| GT2 | 0.1499 | 5.12% | 0.017136 | 0.01133 | 48.30 | 10.3 |
| $\Delta Y_{desired} = 0.03$ | | | | | | |
| T1[10] | 0.1580 | – | 0.003468 | 0.003399 | – | – |
| IT2[10] | 0.1506 | 4.65% | 0.003468 | 0.003502 | 0 | 3.1209 |
| GT2 | 0.1476 | 6.54% | 0.003468 | 0.003502 | 0 | 3.1209 |
| $\Delta Y_{desired} = 0.05$ | | | | | | |
| T1[10] | 0.1588 | – | 0.006732 | 0.9682 | – | – |
| IT2[10] | 0.1508 | 5.09% | 0.008262 | 0.52736 | 45.1452 | -46.895 |
| GT2 | 0.1476 | 7.04% | 0.00816 | 0.527463 | 21.6342 | -46.885 |
| $\Delta Y_{desired} = 0.07$ | | | | | | |
| T1[10] | 0.1711 | – | 0.002142 | 2.932513 | – | – |
| IT2[10] | 0.1595 | 6.80% | 0.006834 | 2.680987 | > 100 | -9.6614 |
| GT2 | 0.1561 | 8.77% | 0.006834 | 2.680884 | > 100 | -8.8374 |
| $\Delta Y_{desired} = 0.1$ | | | | | | |
| T1[10] | 0.1868 | – | 1.03 | 3.3 | – | – |
| IT2[10] | 0.1755 | 6.06% | 1.05 | 3.3 | 0 | 0 |
| GT2 | 0.1721 | 7.86% | 1.08 | 3.3 | 0 | 0 |

The set of 100 input-output pairs is selected randomly as a sample training data. As a multi-input case, the result presented in Table 5.3 shows that GT2 FLS provides improved robust behavior like IT2 FLS. Further, the set of generated input-output data pairs is added by Gaussian noise having zero (0) mean and values of standard deviation 0.1 to the original data. The simulations are again performed by using the same set of GT2 TSK FLS parameters. The obtained results as presented in Table 5.4.

The tabular results demonstrate that GT2 FLS has the capability to provide robust behavior. Though the computation cost associated with GT2 FLS is more in comparison to T1 and IT2 FLS, its advantage for handling a higher level of uncertainty may be useful for various types of applications.

## 5.5 Summary

This chapter has presented a method for robustness investigation of GT2 FLS based on the theoretical background established for IT2 FLS. The GT2 FLS is designed using $\alpha$-plane representation. The robustness analysis for computing the robustness index is modified for $\alpha$-level type-2 FLS. Further, by aggregating all discretized $\alpha$-levels, it is customized for GT2 FLS. The proposed methodology is applied to two function approximation-based simulation examples. The obtained results reveal that, similar to T1 and IT2 FLS, the GT2 FLS can be designed to attain robust behavior with a more flexible structure. The additional design flexibility associated with GT2 FLS provides reduced output error and enhanced robustness performance based on the nature of the application. As GT2 FLS comprises a higher number of design parameters, it can provide more design flexibility in comparison to T1 and IT2 FLS.

## References


1. S. Ovchinnikov, "On robust fuzzy systems," Proceedings of North American Fuzzy Information Processing, 1996, pp. 182–183, https://doi.org/10.1109/NAFIPS.1996.534727.
2. H. T. Nguyen, V. Kreinovich and D. Tolbert, "On robustness of fuzzy logics," [Proceedings 1993] Second IEEE International Conference on Fuzzy Systems, 1993, pp. 543-547 vol.1, https://doi.org/10.1109/FUZZY.1993.327525.
3. W.W. Melek, A.A. Goldenberg, "The development of a robust fuzzy inference mechanism," International Journal of Approximate Reasoning, vol. 39, no. 1, pp. 29–47, 2005.
4. Q. Hu, L. Zhang, S. An, D. Zhang and D. Yu, "On Robust Fuzzy Rough Set Models," in IEEE Transactions on Fuzzy Systems, vol. 20, no. 4, pp. 636–651, 2012, https://doi.org/10.1109/TFUZZ.2011.2181180.
5. Zheng, Zheng & Liu, Wei & Cai, Kai-Yuan, "Robustness of fuzzy operators in environments with random perturbations," Soft Comput., vol. 14, pp. 1339-1348, https://doi.org/10.1007/s00500-009-0497-y
6. Li, Yongming. "Approximation and robustness of fuzzy finite automata." International Journal of Approximate Reasoning, vol. 47, no. 2, pp. 247–257, 2008.
7. Nguyen, Hung T., C-W. Tao, and Wiley E. Thompson. "An empirical study of robustness of fuzzy systems." In [Proceedings 1993] Second IEEE International Conference on Fuzzy Systems, pp. 1340–1345. IEEE, 1993.
8. Wang, Lu, and Keyun Qin. "Robustness of general triple I method for fuzzy soft sets." International Journal of Computational Intelligence Systems, vol. 11, no. 1 (2018):1111, https://doi.org/10.2991/ijcis.11.1.84.
9. Mohammad Biglarbegian, William Melek, Jerry Mendel, "On the robustness of Type-1 and Interval Type-2 fuzzy logic systems in modeling," Information Sciences, Volume 181, Issue 7, 2011, pp. 1325–1347, ISSN 0020-0255.
10. M. Biglarbegian, W. Melek and J. Mendel, "Robustness of interval type-2 fuzzy logic systems," 2010 Annual Meeting of the North American Fuzzy Information Processing Society, 2010, pp. 1–6, https://doi.org/10.1109/NAFIPS.2010.5548209.
11. J. M. Mendel, "General Type-2 Fuzzy Logic Systems Made Simple: A Tutorial," in IEEE Transactions on Fuzzy Systems, vol. 22, no. 5, pp. 1162–1182, 2014, https://doi.org/10.1109/TFUZZ.2013.2286414.
12. J. M. Mendel, F. Liu and D. Zhai, "$\alpha$-Plane Representation for Type-2 Fuzzy Sets: Theory and Applications," in IEEE Transactions on Fuzzy Systems, vol. 17, no. 5, pp. 1189–1207, 2009, https://doi.org/10.1109/TFUZZ.2009.2024411.
13. J. M. Mendel, Uncertain Rule-Based Fuzzy Logic Systems: Introduction and New Directions, NJ, Upper Saddle River: Prentice-Hall, 2001

# Chapter 6
# Interval Type-3 Fuzzy Systems

**Abstract** Using the idea of $\alpha$-plane expression theorems, this chapter provides a design methodology for the development of an Interval Type-3 (IT3) Takagi-Sugeno (T-S) Fuzzy Logic System (FLS). Concept formulation and algorithm development are done using the design baseline of General Type-2 (GT2) FLS and Interval Type-2 (IT2) FLS. As a result, IT3 FLS developed as a replacement for GT2 FLS, with tertiary membership values of unity over the secondary membership Footprint of Uncertainty (FOU) and a Secondary Membership Function (SMF) that is an IT2 fuzzy set. When there is a significant amount of ambiguity and uncertainty, the extra degree of freedom that IT3 FLS offers provides better modeling capabilities than GT2 FLS. When dealing with imprecise data, the proposed system will be more appealing. GT2 and IT2 FS mathematics are used to derive the design methodology and required calculations for implementing the IT3 FLS algorithms. The recommended method is used on a number of benchmark examples, and the actual outcomes are compared to recently created fuzzy modeling techniques. When data is distorted by noise and there are uncertainties related to unvarying or missing data, the projected IT3 T-S FLS model performs well.

## 6.1 Introduction

There is a great deal of ambiguity and uncertainty in the information or data used to describe or analyze the behavior of the system in many real-world applications. These ambiguities could be random (unpredictability) or linguistic (related to words). Klir and Wierman categorized the types of ambiguities as disorder (or strife), imprecision (or non-specificity), and vagueness (or fuzziness) in [1]. According to Berenji [2], in agreement with [1], insufficient accuracy, vagueness, missing information, or randomness can all contribute to uncertainty in data and processes.

Fuzzy logic hypothesis provides a way to deal with both linguistic and spurious ambiguities, and it can function well in the presence of noisy measurements and random disturbances [13]. Type-1 (T1) and type-2 (T2) FLSs are designed using two distinct approaches. In order to address the high level of ambiguity and uncertainty associated with the type-1 FLSs' conception, Zadeh proposed the type-2 FLSs. While

 
D. J. Singh et al., *Fuzzy Systems with Applications to Modeling and Control of an Aircraft*, Algorithms for Intelligent Systems, https://doi.org/10.1007/978-981-95-7888-7_6

T2 FLS MFs are represented by a secondary MF, T1 FLS MFs are clear or definite. The secondary membership in GT2 FLSs is derived from the T1 fuzzy set, which supports greater degrees of freedom, whereas in IT2 FLSs, it is a fixed number [4, 5].

Due to their lower computational cost and structural complexity, the majority of academics favour IT2 FLSs over GT2 FLSs. On the other hand, GT2 FLSs are better at handling noisy and ambiguous data. Mendel demonstrated that GT2 FLSs can perform better than IT2 FLSs in [5, 9]. Additionally, he provided a sufficient illustration of the theoretical shift from type-1 to IT2 to GT2 FLSs, pointing out that the calculations needed for GT2 FLS can be computed using the mathematics of T1 FS and IT2 FS. Liu's $\alpha$-plane representation theorem [6] served as an inspiration for GT2 FLS researchers to simplify and make it more applicable [7, 11]. The additional properties of GT2 Fuzzy Sets (FSs) were also investigated in some studies. Kulkarni et al. [10] proposes a novel approach that adaptively determines the optimal fuzzifier values for the $\alpha$-planes of GT2 FSs. References [52, 63, 64] discuss the idea of type-2 intuitionistic FSs. Wu and Mendel have presented a thorough examination of Jaccard similarity metrics for GT2 FSs in [15]. The constrained type-2 (CT2) fuzzy sets, which have limitations on the shape of the embedded sets and footprint of uncertainty, are introduced in [55]. In [56], the concept of thick fuzzy sets (TFSs) and their applicability for uncertain fuzzy modeling and computations are examined. Chen et al. [57] offers a comprehensive analysis of the effectiveness of two new type-reduction algorithms, as well as the ones that are currently in use.

Over the past few years, GT2 FLSs have been employed in numerous engineering applications. For instance, design of fuzzy controllers [16–18], brain-machine communication system of a humanoid robot [19, 20], to address bearing defect detection issues [21]. A detailed examination of T2 FLS applications is given in [52] with noise endurance study presented in [53]. Ontiveros et al. [54] describes a novel application of Type-2 FLS for medical diagnostics. These studies show that GT2 FLSs can handle uncertainty and ambiguity better than IT2 FLSs.

It has been suggested that a relation can be made more fuzzy by using higher-order. "More fuzziness in the description denotes a greater capacity to rationally handle imprecise information," according to Hisdal [3]. The grey area is left open for further research because higher-type fuzzy sets (type-3 or higher) are rarely the focus of research in the literature. The notion of fuzzy subset hood for generalized type-n fuzzy sets was developed by Rickard et al. in [23, 45]. A succinct review of type-3 or higher-order fuzzy sets is provided in [24, 25, 51] as a potential area of future study in fuzzy logic theory. Castillo presented a method in [25] for determining the ideal value of n in type-n fuzzy systems for specific problem classes. The idea of hesitant fuzzy sets and a synopsis of type-3 and type-n FSs are introduced in [26, 27]. The fundamental concept is emphasized in [27]. Aisbett et al. proposed the set theory operations for type-3 and higher type fuzzy sets in [44]. Fisher et al. investigated the higher-order ambiguity for type-n fuzzy sets in [46]. Turksen claimed in [47] that an iterative method outlined for type-2 FS could be used to construct type-3 to type-n FS.

Mohammadzadeh et al. recently developed the baseline for interval type-3 FLS in [22]. A novel control strategy is presented in [48] for a class of uncertain nonlinear systems. Liu et al. in [49] design the IT3 fuzzy control system for managing solar energy using the theoretical foundation provided in [22]. Additionally, in [50], Ma et al. solved singular multi-pantograph equations using the previously mentioned type-3 FLS. Motivated by these endeavours, the authors proposed a systematic method for developing interval type-3 T-S fuzzy systems based on alpha plane representation theorems.

The IT2 FS [4] and GT2 FS [5] mathematics are used to derive the computations needed for the IT3 FLS design. The baseline IT2 T-S fuzzy model was designed using the Hyper-Plane-Shaped Membership Function (HPSMF) and the modified interval type-2 FCRM clustering algorithm [29]. The mathematical formulas in [29–34] have helped to initiate the foundational work for this study.

## 6.2 Brief Review on Interval Type-2 T-S Fuzzy System

This section provides a concise theoretical background on GT2 fuzzy sets (FSs) along with their $\alpha$-plane representation. For completeness, the mathematical relations from [5, 7] are also restated.

In GT2 FSs, the membership function (MF) is three-dimensional, where each primary MF is associated with a corresponding secondary MF. The generalized form of a type-2 fuzzy set $\tilde{\mathbf{F}}$ is expressed as

$$\tilde{F} = \{((x, u), \mu_{\tilde{F}}(x, u)) \mid x \in X, \ u \in [0, 1]\} \tag{6.1}$$

The secondary MF depends on both the primary and secondary variables, and is written as $\mu_{\tilde{F}}(x, u)$, where $0 \leq \mu_{\tilde{F}}(x, u) \leq 1$. An alternative representation of type-2 FS is given as

$$\tilde{F} = \int_{x \in X} \int_{u \in [0,1]} \mu_{\tilde{F}}(x, u)/(x, u) \tag{6.2}$$

where $\int \int$ represents union over all permissible values of $x$ and $u$.

Fuzzy set $\tilde{F}$ is called an IT2 FS when $\mu_{\tilde{F}}(x, u) = 1, \forall x \in X, \forall u \in J_x$, it is commonly represented as

$$\tilde{F} = \int_{x \in X} \int_{u \in [0,1]} 1/(x, u) \tag{6.3}$$

The Footprint Of Uncertainty (FOU) of $\tilde{F}$ is the 2-D support of $\mu_{\tilde{F}}(x, u)$ and is characterized as:

$$\text{FOU}(\tilde{F}) = \{(x, u) \in X \times [0, 1] \mid \mu_{\tilde{F}}(x, u) > 0\} \tag{6.4}$$

The FOU $\tilde{F}$ bounded by Upper Membership Function (UMF) and Lower Membership Function (LMF) is described as:

$$\text{LMF}(\tilde{F}) = \underline{\mu}_{\tilde{F}}(x) = \inf\{u \mid u \in [0, 1], \mu_{\tilde{F}}(x, u) > 0\} \tag{6.5}$$

$$\text{UMF}(\tilde{F}) = \overline{\mu}_{\tilde{F}}(x) = \sup\{u \mid u \in [0, 1], \mu_{\tilde{F}}(x, u) > 0\} \tag{6.6}$$

At each $x \in X$ of $\tilde{F}$, the primary membership $J_x$ is:

$$J_x = \{u \in [0, 1] \mid \mu_{\tilde{F}}(x, u) > 0\} = [\underline{\mu}_{\tilde{F}}(x), \overline{\mu}_{\tilde{F}}(x)] \tag{6.7}$$

The MF of a GT2 FS, defined in 3-D space, can be decomposed using horizontal, vertical, or wavy slices.

The vertical slice expression is represented as follows

Let $\tilde{F}_\alpha(x)$ denote the $\alpha$-cut of the type-1 FS $\tilde{F}(x)$ for $\alpha \in [0, 1]$, then

$$\tilde{F}_\alpha(x) = \{u \mid \mu_{\tilde{F}(x)}(u) \geq \alpha\} = [a_\alpha(x), b_\alpha(x)], \quad \alpha \in [0, 1] \tag{6.8}$$

For every $x \in X$, the $\alpha$-cut decomposition of $\tilde{F}(x)$ is

$$\tilde{F}(x) = \bigcup_{\forall\alpha\in[0,1]} [\alpha/\tilde{F}_\alpha(x)] = \sup_{\forall\alpha\in[0,1]} [\alpha/\tilde{F}_\alpha(x)] = \sup_{\forall\alpha\in[0,1]} [\alpha/[a_\alpha(x), b_\alpha(x)]] \tag{6.9}$$

$$\tilde{F} = \int_{x\in X} \left[ \bigcup_{\forall\alpha\in[0,1]} [\alpha/\tilde{F}_\alpha(x)] \right] \Big/ x \tag{6.10}$$

Equation (6.10) gives the vertical slice of $\tilde{F}$. For each $x$, a type-1 secondary FS exists, defined through its $\alpha$-cut decomposition [14]. Furthermore, Eq. (6.10) can be rewritten as

$$\tilde{F} = \bigcup_{\forall\alpha\in[0,1]} \left[ \alpha/\left( \int_{\forall x\in X} [a_\alpha(x), b_\alpha(x)]/x \right) \right] \tag{6.11}$$

From (6.11), the plane at level $\alpha$ is denoted by

$$\alpha \left[ \int_{\forall x\in X} [a_\alpha(x), b_\alpha(x)]/x \right].$$

The $\alpha$-plane obtained from $\alpha$-cuts of secondary membership functions will be

$$\tilde{F}_\alpha = \int_{\forall x\in X} \tilde{F}_\alpha(x)/x = \int_{\forall x\in X} [a_\alpha(x), b_\alpha(x)]/x \tag{6.12}$$

## 6.3 Interval Type-3 T-S Fuzzy Set

The notion of interval type-3 FS is developed using $\alpha$-plane representation for type-2 fuzzy sets and Zadeh's definition for type-$n$ fuzzy sets. Turksen has projected the type-3 fuzzy set membership function and explained that a type-3 fuzzy set is created by combining two complete type-2 FSs in [28]. Additionally, he pointed out that the probabilistic system with first, second, and third moments may be thought of as being similar to type-1, type-2, and type-3 FSs. Torra and Narukawa noted in [27] that type-3 FSs allow type-2 FSs to be members. In [23, 45], Rickard et al. conceptualized about type-$n$ fuzzy sets, in the context of its subset hood as:

For $n > 1$, type-$n$ FSs are the sets whose MF takes values in type $(n-1)$ FSs on the unit interval.

In generalized form, let $\tilde{F}^{(n)}$ signify a type-$n$ FS for $n > 1$ and $\gamma_{n-1}$ is a set of type-$(n-1)$ FSs on unit interval, or equivalently, the set of secondary MFs $f : [0, 1]^{n-1} \to [0, 1]$. Then, the membership function (MF) of a type-n fuzzy set $\tilde{F}^{(n)}$, is a measurable mapping $\mu_{\tilde{F}^{(n)}} : X \to \gamma_{n-1}$ termed as the primary MF. Hence, the fuzzy set $\tilde{F}^{(n)}$ can also be recognized by the mapping $\mu_{\tilde{F}^{(n)}} : (X \times [0, 1]^{n-1} \to [0, 1]) \to \gamma_1$.

In general, a type-$n$ fuzzy set $\tilde{F}^{(n)}$ is defined by its primary MF $\mu_{\tilde{F}^{(n)}} : (X \times [0, 1]^{n-2}) \to \gamma_1$. Here, a type-$n$ fuzzy set on $X$ can be interpreted as a type-1 fuzzy set on $(X \times [0, 1]^{n-2})$. From this perspective, interval type-2 fuzzy sets are extended to interval type-$n$ fuzzy sets.

A type-3 fuzzy set can be seen as an advancement of the general type-2 fuzzy set, where the secondary MF itself is an IT2 fuzzy set. The tertiary membership values are given by $\mu_{\tilde{F}^{(3)}} : X \times [0, 1]^2 \to [0, 1]$, and remain equal to unity within the footprint of uncertainty (FOU) bounded by the lower and upper membership functions $X \times [0, 1] \to [0, 1]$.

The formulation of an IT3 fuzzy set (Fig. 6.1) is derived from [6].

In [6, Definition1], $\alpha$-plane is,

$$\tilde{F}_{\alpha} = \bigcup_{x \in X} (x, u) \mid \mu_{\tilde{F}}(x, u) \geq \alpha = \bigcup_{x \in X} \left(\mu_{\tilde{F}}(x)\right)_{\alpha} \tag{6.13}$$

If the secondary grade $\alpha$ becomes an interval set as

$$\tilde{\alpha} \in [\underline{\alpha}, \overline{\alpha}],$$

Then, Definition 1 can be expressed as

$$\tilde{F}_{\tilde{\alpha}} = \bigcup_{x \in X} (x, u) \mid \underline{\alpha} \leq \mu_{\tilde{F}}(x, u) < \overline{\alpha} = \bigcup_{x \in X} \left(\mu_{\tilde{F}}(x)\right)_{\tilde{\alpha}} \tag{6.14}$$

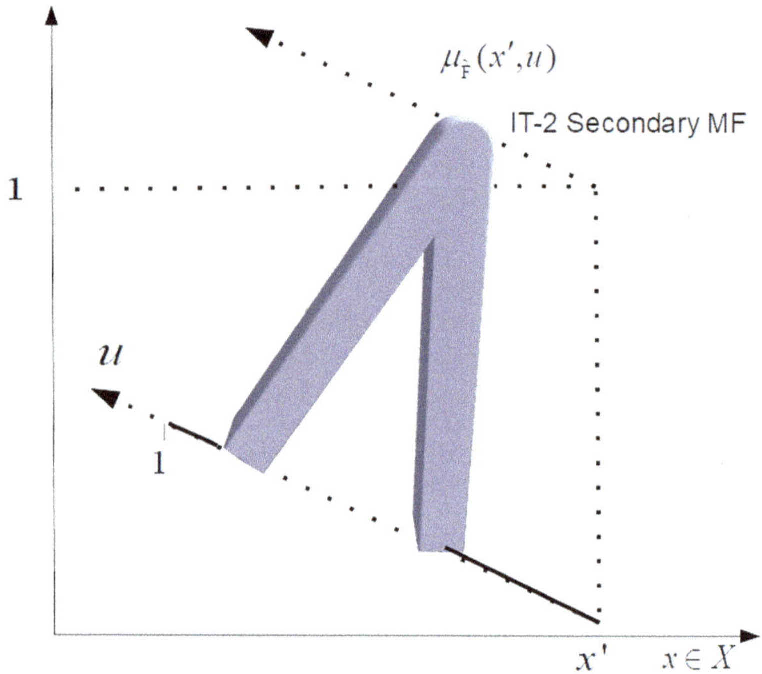


**Fig. 6.1** Illustration of IT3 FSs (Rickard et al. [23], 2009, IEEE TFS)

Similarly, [6, Definition 2] for associated type-2 fuzzy set of the $\alpha$-plane $\tilde{F}_\alpha$ is defined as

$$\tilde{F}(\alpha) = \{(x, u), \alpha \mid \forall (x, u) \in \tilde{F}_\alpha\}. \tag{6.15}$$

Equation (6.15) can be rewritten for interval secondary grade $\alpha$ as

$$\tilde{F}(\tilde{\alpha}) = \{(x, u), [\underline{\alpha}, \overline{\alpha}] \mid \forall (x, u) \in \tilde{F}_{\tilde{\alpha}}\}. \tag{6.16}$$

Mendel [8] suggested that $\alpha$-level T2 FSs can be written as,

$$R_{\tilde{F}_\alpha}(x, u) = \frac{\alpha}{\tilde{F}_\alpha}(\forall x \in X, u \in J_x)$$

where

If $\alpha$ plane is interval set, then $\alpha$-level T2 FSs would be written as "uncertain $\alpha$-level T2 FSs".

Inspiring with $\alpha$-level T2 FSs $R_{\tilde{F}_\alpha}(x, u)$, the $\alpha$ -level interval T3 FSs with interval secondary membership grade $[\underline{\alpha}, \overline{\alpha}]$ and unity tertiary membership grade are introduced as

$$R_{\tilde{F}_\alpha}(x, u) = \frac{1/\tilde{\alpha}}{\tilde{F}_{\tilde{\alpha}}}, \forall x \in X, u \in J_x, \ \tilde{\alpha} \subseteq [\underline{\alpha}, \overline{\alpha}] \tag{6.17}$$

where $\tilde{\alpha} \subseteq [\underline{\alpha}, \overline{\alpha}]$, and corresponding to $\underline{\alpha}$ and $\overline{\alpha}$ there are $\tilde{F}_{\underline{\alpha}} \subseteq [\underline{F}_{\underline{\alpha}}, \overline{F}_{\underline{\alpha}}]$ and $\tilde{F}_{\overline{\alpha}} \subseteq [\underline{F}_{\overline{\alpha}}, \overline{F}_{\overline{\alpha}}]$ respectively.

Equation 6.17 can also be rewritten as

$$R_{\tilde{F}_{\alpha}}(x,u) = \left[ \frac{1/\underline{\alpha}}{[\underline{F}_{\underline{\alpha}}, \overline{F}_{\underline{\alpha}}]}, \frac{1/\overline{\alpha}}{[\underline{F}_{\overline{\alpha}}, \overline{F}_{\overline{\alpha}}]} \right], \quad \forall x \in X, u \in J_x \tag{6.18}$$

If $\tilde{\alpha}_{\lambda},\ \lambda = 1, 2, \ldots, \rho$ represent the interval secondary grades, i.e. $\tilde{\alpha}_{\lambda} \subseteq [\underline{\alpha}_{\lambda}, \overline{\alpha}_{\lambda}]$, then at each level $\tilde{\alpha}_{\lambda}$, $\tilde{F}_{\tilde{\alpha}_{\lambda}}$, the horizontal slice representation of $\tilde{F}$ is as (6.19)

$$\tilde{F} = \frac{1/\tilde{\alpha}_1}{\sum_{x \in X} \tilde{F}_{\tilde{\alpha}_1}/x} + \frac{1/\tilde{\alpha}_2}{\sum_{x \in X} \tilde{F}_{\tilde{\alpha}_2}/x} + \cdots + \frac{1/\tilde{\alpha}_{\rho}}{\sum_{x \in X} \tilde{F}_{\tilde{\alpha}_{\rho}}/x} \tag{6.19}$$

Equation (6.19) can be further represented as (6.19)

$$\tilde{F} = \left[ \frac{1/\underline{\alpha}_1}{\sum_{x \in X} \frac{\underline{F}_{\underline{\alpha}_1}, \overline{F}_{\underline{\alpha}_1}}{x}}, \frac{1/\overline{\alpha}_1}{\sum_{x \in X} \frac{\underline{F}_{\overline{\alpha}_1}, \overline{F}_{\overline{\alpha}_1}}{x}} \right] + \cdots + \left[ \frac{1/\underline{\alpha}_{\rho}}{\sum_{x \in X} \frac{\underline{F}_{\underline{\alpha}_{\rho}}, \overline{F}_{\underline{\alpha}_{\rho}}}{x}}, \frac{1/\overline{\alpha}_{\rho}}{\sum_{x \in X} \frac{\underline{F}_{\overline{\alpha}_{\rho}}, \overline{F}_{\overline{\alpha}_{\rho}}}{x}} \right] \tag{6.20}$$

The continuous IT3 FS is represented as (6.21)

$$\tilde{F} = \int_{x \in X} \int_{u \in [0,1]} \left[ \frac{1}{\mu_{\tilde{F}}(x,u)} \right] \cdot \frac{1}{(x,u)} \tag{6.21}$$

Here $\int \int$ signifies union and tertiary membership grade is unity [22].

## 6.4 Design Methodology for Interval Type-3 T-S Fuzzy System

This section outlines the design of the IT3 T-S fuzzy logic system based on the IT2 secondary MF concept. The role of design parameters and computational complexity is highlighted. The framework builds on GT2 and IT2 FLS, where modified IT2 FCRM clustering with HPSMF is employed to form the IT2 T-S FLS structure [29].

Assume that data pairs $(\mathbf{x}_k, y_k)$ for $k = 1, 2, ...., N$ are arranged in $c$ clusters and $i$th cluster is in form of hyper-plane regression function as

$$\hat{y}_k = f^i(\mathbf{x}_k, \boldsymbol{\omega}_i) = a_1^i x_{k1} + a_2^i x_{k2} + \cdots + a_M^i x_{kM} + a_0^i = \left[\mathbf{x}_k\ 1\right] \cdot \boldsymbol{\omega}_i^T \tag{6.22}$$

where

$$\mathbf{x}_k = [x_{k1}, x_{k2}, \ldots, x_{kM}]$$

is the $k$th input vector, $i = 1, 2, \ldots, c$ is the number of clusters, and $j = 1, 2, \ldots, M$ is the number of input variables in each regression.

The hyperplane model coefficient vector is

$$\boldsymbol{\omega}_i = [a_1^i, a_2^i, \ldots, a_M^i, a_0^i].$$

To formulate the IT2 fuzzy set, two fuzzifiers are considered with the objective function corresponding to two fuzzy exponents $m_1$ and $m_2$ are (6.23) and (6.24)

$$\begin{cases} \min J_{m_1}(\mathbf{U}, \boldsymbol{\omega}) = \sum_{k=1}^{N} \sum_{i=1}^{c} (\mu_{ik})^{m_1} E_{ik}(\omega_i), \\ \sum_{i=1}^{c} u_{ik} = 1, \quad k = 1, 2, \ldots, N \end{cases} \tag{6.23}$$

$$\begin{cases} \min J_{m_2}(\mathbf{U}, \boldsymbol{\omega}) = \sum_{k=1}^{N} \sum_{i=1}^{c} (\mu_{ik})^{m_2} E_{ik}(\omega_i), \\ \sum_{i=1}^{c} u_{ik} = 1, \quad k = 1, 2, \ldots, N \end{cases} \tag{6.24}$$

The partition matrix is $\mathbf{U} = [u_{ik}]_{c \times N}$. The parameter $E_{ik}$ is the error distance between the estimated regression equation and actual output [41].

$$E_{ik}(\boldsymbol{\omega}_i) = \left(y_k - f^i(\mathbf{x}_k, \boldsymbol{\omega}_i)\right)^2, \quad i = 1, 2, \ldots, c, \; k = 1, 2, \ldots, N \tag{6.25}$$

The upper membership $\overline{u}_{ik}$ and lower membership $\underline{u}_{ik}$ are obtained by Lagrange multiplier as

$$\overline{u}_{ik} = \begin{cases} \dfrac{1}{\sum_{r=1}^{c} \left[\dfrac{E_{ik}(\omega_i)}{E_{rk}(\omega_i)}\right]^{\frac{2}{m_1 - 1}}}, & \dfrac{1}{\sum_{r=1}^{c} \left[\dfrac{E_{ik}(\omega_i)}{E_{rk}(\omega_i)}\right]} < \dfrac{1}{c}, \\ \dfrac{1}{\sum_{r=1}^{c} \left[\dfrac{E_{ik}(\omega_i)}{E_{rk}(\omega_i)}\right]^{\frac{2}{m_2 - 1}}}, & \text{otherwise.} \end{cases} \tag{6.26}$$

$$\underline{u}_{ik} = \begin{cases} \dfrac{1}{\sum_{r=1}^{c} \left[\dfrac{E_{ik}(\omega_i)}{E_{rk}(\omega_i)}\right]^{\frac{2}{m_1 - 1}}}, & \dfrac{1}{\sum_{r=1}^{c} \left[\dfrac{E_{ik}(\omega_i)}{E_{rk}(\omega_i)}\right]} \geq \dfrac{1}{c}, \\ \dfrac{1}{\sum_{r=1}^{c} \left[\dfrac{E_{ik}(\omega_i)}{E_{rk}(\omega_i)}\right]^{\frac{2}{m_2 - 1}}}, & \text{otherwise.} \end{cases} \tag{6.27}$$

Based on the above algorithms, the data is partitioned into hyperplane-shaped clusters. The updated membership function matrices comprising membership degree are obtained as $\bar{\mathbf{U}} = [\bar{u}_{ik}]_{c\times N}$ and $\underline{\mathbf{U}} = [\underline{u}_{ik}]_{c\times N}$. The upper and lower hyperplanes obtained by Weighted Least Squares (WLS) are as

$$\bar{\omega}_i = \left[\mathbf{X}^T \bar{\mathbf{P}}_i \mathbf{X}\right]^{-1} \mathbf{X}^T \bar{\mathbf{P}}_i \mathbf{y} \tag{6.28}$$

$$\underline{\omega}_i = \left[\mathbf{X}^T \underline{\mathbf{P}}_i \mathbf{X}\right]^{-1} \mathbf{X}^T \underline{\mathbf{P}}_i \mathbf{y} \tag{6.29}$$

where

$$\mathbf{X} = \left[\mathbf{x}_k \ \ 1\right]_{N\times(M+1)} \quad \text{and} \quad \mathbf{y} = \left[y_k\right]_{N\times 1}$$

The partition matrix $\bar{\mathbf{P}}_i$ and $\underline{\mathbf{P}}_i$ are as

$$\bar{\mathbf{P}}_i = \begin{bmatrix} \bar{u}_{i1} & 0 & \cdots & 0 \\ 0 & \bar{u}_{i2} & \cdots & 0 \\ \vdots & \vdots & \ddots & \vdots \\ 0 & 0 & \cdots & \bar{u}_{iN} \end{bmatrix}, \quad \underline{\mathbf{P}}_i = \begin{bmatrix} \underline{u}_{i1} & 0 & \cdots & 0 \\ 0 & \underline{u}_{i2} & \cdots & 0 \\ \vdots & \vdots & \ddots & \vdots \\ 0 & 0 & \cdots & \underline{u}_{iN} \end{bmatrix} \in R^{N\times N}.$$

The upper and lower membership degrees calculated using hyperplane-shaped Gaussian MF are as follows:

$$\bar{\mu}_{\tilde{F}_i}(\mathbf{x}_k) = \exp\left(-\eta \cdot \frac{d_{ik}(\bar{\omega}_i)}{\max\left(\{d_{ik}(\bar{\omega}_i),\ i = 1, \ldots, c\}\right)}\right) \tag{6.30}$$

$$\underline{\mu}_{\tilde{F}_i}(\mathbf{x}_k) = \exp\left(-\eta \cdot \frac{d_{ik}(\underline{\omega}_i)}{\max\left(\{d_{ik}(\underline{\omega}_i),\ i = 1, \ldots, c\}\right)}\right) \tag{6.31}$$

where $d_{ik}(\bar{\omega}_i)$ and $d_{ik}(\underline{\omega}_i)$, the distance of $\mathbf{x}_k$ from $i$th upper and lower hyper planes are expressed as

$$d_{ik}(\bar{\omega}_i) = \frac{|\mathbf{x}_k \cdot \bar{\omega}_i|}{\|\bar{\omega}_i\|} \quad \text{and} \quad d_{ik}(\underline{\omega}_i) = \frac{|\mathbf{x}_k \cdot \underline{\omega}_i|}{\|\underline{\omega}_i\|}$$

For an IT2 T-S FLS, $i$th rule is written as,

$$R^i : \text{IF } x_1 \text{ is } \tilde{F}_1^i \text{ and } x_2 \text{ is } \tilde{F}_2^i \text{ and } \ldots \text{ and } x_p \text{ is } \tilde{F}_p^i$$

$$\text{THEN } y^i(\mathbf{x}) = C_1^i x_1 + C_2^i x_2 + \cdots + C_M^i x_M + C_0^i \tag{6.32}$$

The consequent parameters ($C_j^i$ and $C_0^i$) are represented as [13]:

$$C_j^i = \left[v_j^i - s_j^i,\ v_j^i + s_j^i\right],\quad C_0^i = \left[v_0^i - s_0^i,\ v_0^i + s_0^i\right]$$

where $\boldsymbol{v}_j^i$ signifies the mean and $s_j^i$ the spread of $C_j^i$ and are obtained from the coefficients of the upper and lower hyperplanes using

$$\bar{\boldsymbol{\omega}}_i = \left(\bar{a}_1^i, \bar{a}_2^i, \ldots, \bar{a}_M^i, \bar{a}_0^i\right),\quad \underline{\boldsymbol{\omega}}_i = \left(\underline{a}_1^i, \underline{a}_2^i, \ldots, \underline{a}_M^i, \underline{a}_0^i\right)$$

as follows:

$$n_j^i = (\bar{a}_j^i + \underline{a}_j^i)/2,\quad n_0^i = (\bar{a}_0^i + \underline{a}_0^i)/2,$$

and

$$s_j^i = (\bar{a}_j^i - \underline{a}_j^i)/2,\quad s_0^i = (\bar{a}_0^i - \underline{a}_0^i)/2$$

Hence, consequent of rule $R^i$ is expressed as $Y^i = [y_l^i, y_r^i]$ where

$$y_l^i = \sum_{k=1}^{p} v_j^i \mathbf{x}_k + v_0^i - \sum_{k=1}^{p} s_j^i |\mathbf{x}_k| - s_0^i \tag{6.33}$$

$$y_r^i = \sum_{k=1}^{p} v_j^i \mathbf{x}_k + v_0^i + \sum_{k=1}^{p} s_j^i |\mathbf{x}_k| + s_0^i \tag{6.34}$$

The base of each triangular type secondary MF is $\overline{\mu}_{\tilde{F}}(x) - \underline{\mu}_{\tilde{F}}(x)$ with apex position [5] as,

$$\text{apex}^{(x) = \underline{\mu}_{\tilde{F}}(x) + w\left[\bar{\mu}_{\tilde{F}}(x) - \underline{\mu}_{\tilde{F}}(x)\right]}, \tag{6.35}$$

The $\alpha$-cuts of triangle secondary MFs [5, 9] are as

$$\begin{cases} \tilde{F}_\alpha(x) = \left[a_\alpha(x),\ b_\alpha(x)\right], \\ a_\alpha(x) = \mu_{\underline{F}}(x) + w\left[\mu_{\overline{F}}(x) - \mu_{\underline{F}}(x)\right]\alpha, \\ b_\alpha(x) = \mu_{\overline{F}}(x) - (1 - w)\left[\mu_{\overline{F}}(x) - \mu_{\underline{F}}(x)\right]\alpha. \end{cases} \tag{6.36}$$

For each rule, the $\alpha$-cut decomposition of type-1 vertical slice of rule-antecedent (when $x_j = x_j'$) $\tilde{F}_j^i(x_j')$ is obtained as

$$\tilde{F}_j^i(x_j') = \sup_{\forall \alpha \in [0,1]} \alpha / \left[a_{j,\alpha}^i(x_j'),\ b_{j,\alpha}^i(x_j')\right] \tag{6.37}$$

For level $\alpha$, the firing interval $\tilde{F}_\alpha^i(x')$ is as

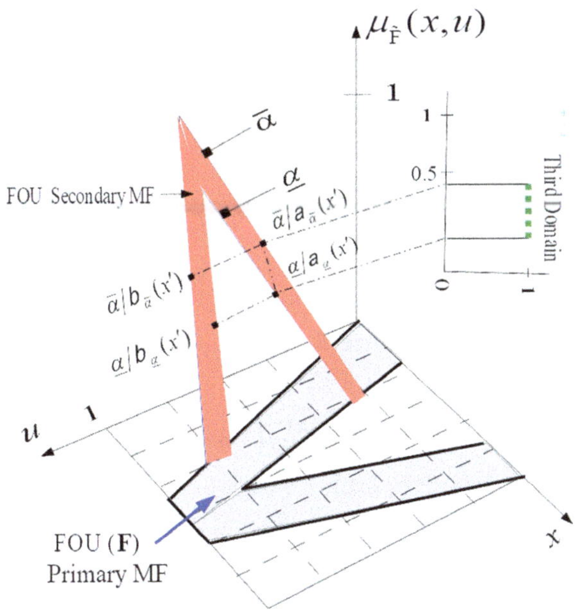


Fig. 6.2 Secondary MF of IT3 FSs

$$F_\alpha : \begin{cases} \tilde{F}^i_\alpha(x') \equiv \left[\, \underline{f}^i_\alpha(x'),\ \overline{f}^i_\alpha(x') \,\right], \\ \underline{f}^i_\alpha(x') = \prod_{j=1}^{p} a^i_{j,\alpha}(x'_j), \\ \overline{f}^i_\alpha(x') \equiv \prod_{j=1}^{p} b^i_{j,\alpha}(x'_j). \end{cases} \qquad (6.38)$$

where $T$ denotes a t-norm, either the product or minimum.

Similar to how the baseline of T1 FLS was used for the design of IT2 FLS, the baseline of GT2 FLS is used for the design of IT3 FLS. Information about the Footprint of Uncertainty (FOU) of primary MFs is used in the design of IT2 FLS. This design feature encourages us to design the IT3 FLS using the FOU of secondary MF.

The design of IT3 FS also makes use of the representation of the $baralpha$ and $underlinealpha$ planes of GT2 FS as detailed in Sect. 6.4.

Figure 6.2 shows the FOU for Secondary MF (SMF) of IT3 FS. The cut points on lower SMF ($\underline{\alpha}$) and upper SMF ($\bar{\alpha}$) are represented as $\underline{\alpha}/\left[a_{\underline{\alpha}}(x'),\ b_{\underline{\alpha}}(x')\right]$ and $\bar{\alpha}/\left[a_{\bar{\alpha}}(x'),\ b_{\bar{\alpha}}(x')\right]$ respectively. The tertiary membership grade is unity.

Each $\bar{\alpha}$ and $\underline{\alpha}$ are discretize for $\lambda = 1, ...., \rho$ levels. Then the IT3 MF are generated by taking the union of all horizontal slices $\tilde{\alpha_1}, \tilde{\alpha_2}, ...., \tilde{\alpha_\lambda}$, where $\tilde{\alpha}_\lambda \subseteq [\underline{\alpha}_\lambda, \bar{\alpha}_\lambda]$. The horizontal slice of IT3 MF at level $\tilde{\alpha_\lambda}$ signifies an interval type-2 MF where secondary grade is between $[\underline{\alpha}_\lambda, \bar{\alpha}_\lambda]$.

The step-by-step procedure for the design of IT3 T-S FLS is as follows:

1. Select the $\alpha$-cuts number $\rho$ for all the vertical slices of $\tilde{F}$ and discretize the values of $\alpha_\lambda$ as $\alpha_1, \alpha_2, \ldots, \alpha_\rho$ for $\lambda = 1, \ldots, \rho$ levels.
   Represent the uncertainty in secondary grades as

$$[\underline{\alpha}_\lambda,\ \bar{\alpha}_\lambda]$$

   where $\underline{\alpha}_\lambda = \alpha_\lambda^\Delta$ and $\bar{\alpha}_\lambda = \alpha_\lambda^{1/\Delta}$, i.e., $\underline{\alpha}_\lambda$ as $\underline{\alpha}_1, \underline{\alpha}_2, \ldots, \underline{\alpha}_\rho$ and $\bar{\alpha}_\lambda$ as $\bar{\alpha}_1, \bar{\alpha}_2, \ldots, \bar{\alpha}_\rho$.
   The fuzziness parameter $\Delta \geq 1$ is a constant and represents the uncertainty of the vertical slice at $x_i = x_i'$.
2. For each discretize $\mathbf{X}$ (denoted as $x_k$, for $k = 1, \ldots, N$) over the support of $\tilde{F}$, compute

$$\left[a^i_{\underline{\alpha}_\lambda}(\mathbf{x}_k),\ b^i_{\underline{\alpha}_\lambda}(\mathbf{x}_k)\right] \text{ and } \left[a^i_{\bar{\alpha}_\lambda}(\mathbf{x}_k),\ b^i_{\bar{\alpha}_\lambda}(\mathbf{x}_k)\right] \text{ for } \lambda = 1, \ldots, \rho$$

   as

$$\begin{cases} a^i_{\underline{\alpha}_\lambda}(x_k) = \mu_{\underline{F}_i}(x_k) + w\left[\mu_{\overline{F}_i}(x_k) - \mu_{\underline{F}_i}(x_k)\right]\underline{\alpha}_\lambda, \\ b^i_{\underline{\alpha}_\lambda}(x_k) = \mu_{\overline{F}_i}(x_k) - (1-w)\left[\mu_{\overline{F}_i}(x_k) - \mu_{\underline{F}_i}(x_k)\right]\underline{\alpha}_\lambda. \end{cases} \tag{6.39}$$

$$\begin{cases} a^i_{\bar{\alpha}_\lambda}(x_k) = \mu_{\underline{F}_i}(x_k) + w\left[\mu_{\overline{F}_i}(x_k) - \mu_{\underline{F}_i}(x_k)\right]\bar{\alpha}_\lambda, \\ b^i_{\bar{\alpha}_\lambda}(x_k) = \mu_{\overline{F}_i}(x_k) - (1-w)\left[\mu_{\overline{F}_i}(x_k) - \mu_{\underline{F}_i}(x_k)\right]\bar{\alpha}_\lambda. \end{cases} \tag{6.40}$$

3. Obtain the firing interval $\tilde{F}^i_{\underline{\alpha}_\lambda}(\mathbf{x}_k)$ and $\tilde{F}^i_{\bar{\alpha}_\lambda}(\mathbf{x}_k)$ for each level $\underline{\alpha}_\lambda$ and $\bar{\alpha}_\lambda$ corresponding to the $i$th rule as

$$\begin{cases} \tilde{F}^i_{\underline{\alpha}_\lambda}(x_k) \equiv \left[\underline{f}^i_{\underline{\alpha}_\lambda}(x_k),\ \overline{f}^i_{\underline{\alpha}_\lambda}(x_k)\right], \\ \tilde{F}^i_{\bar{\alpha}_\lambda}(x_k) \equiv \left[\underline{f}^i_{\bar{\alpha}_\lambda}(x_k),\ \overline{f}^i_{\bar{\alpha}_\lambda}(x_k)\right]. \end{cases} \tag{6.41}$$

   where,

$$\left[\underline{f}^i_{\underline{\alpha}_\lambda}(\mathbf{x}_k),\ \overline{f}^i_{\underline{\alpha}_\lambda}(\mathbf{x}_k)\right] \equiv \left[a^i_{\underline{\alpha}_\lambda}(\mathbf{x}_k),\ b^i_{\underline{\alpha}_\lambda}(\mathbf{x}_k)\right]$$

   and

$$\left[\underline{f}^i_{\bar{\alpha}_\lambda}(\mathbf{x}_k),\ \overline{f}^i_{\bar{\alpha}_\lambda}(\mathbf{x}_k)\right] \equiv \left[a^i_{\bar{\alpha}_\lambda}(\mathbf{x}_k),\ b^i_{\bar{\alpha}_\lambda}(\mathbf{x}_k)\right]$$

4. Compute $\underline{\alpha}_\lambda$ and $\bar{\alpha}_\lambda$ level $m-n$ outputs

$$y_{mn,\underline{\alpha}_\lambda}(\mathbf{x}_k) \text{ and } y_{mn,\bar{\alpha}_\lambda}(\mathbf{x}_k)$$

   at each $\underline{\alpha}_1, \underline{\alpha}_2, \ldots, \underline{\alpha}_\rho$ and $\bar{\alpha}_1, \bar{\alpha}_2, \ldots, \bar{\alpha}_\rho$ as [5, 12].

$$y_{mn,\underline{\alpha}_\lambda}(\mathbf{x}_k) \equiv m\frac{\sum_{i=1}^{c} y_l^i \underline{f}_{\underline{\alpha}_\lambda}^i(\mathbf{x}_k)}{\sum_{i=1}^{c} \underline{f}_{\underline{\alpha}_\lambda}^i(\mathbf{x}_k)} + n\frac{\sum_{i=1}^{c} y_r^i \bar{f}_{\underline{\alpha}_\lambda}^i(\mathbf{x}_k)}{\sum_{i=1}^{c} \bar{f}_{\underline{\alpha}_\lambda}^i(\mathbf{x}_k)} \tag{6.42}$$

$$y_{mn,\bar{\alpha}_\lambda}(x_k) \equiv m\frac{\sum_{i=1}^{c} y_l^i \underline{f}_{\bar{\alpha}_\lambda}^i(\mathbf{x}_k)}{\sum_{i=1}^{c} \underline{f}_{\bar{\alpha}_\lambda}^i(\mathbf{x}_k)} + n\frac{\sum_{i=1}^{c} y_r^i \bar{f}_{\bar{\alpha}_\lambda}^i(\mathbf{x}_k)}{\sum_{i=1}^{c} \bar{f}_{\bar{\alpha}_\lambda}^i(\mathbf{x}_k)} \tag{6.43}$$

where consequent of rule $R^i$, i.e. $[v_j^i, v_r^i]$ are obtained from (6.33)–(6.34).

5. Defuzzify the $y_{mn,\underline{\alpha}_\lambda}(\mathbf{x}_k)$ and $y_{mn,\bar{\alpha}_\lambda}(\mathbf{x}_k)$ for $\underline{\alpha}_1, \underline{\alpha}_2, \ldots, \underline{\alpha}_\rho$ and $\bar{\alpha}_1, \bar{\alpha}_2, \ldots, \bar{\alpha}_\rho$, and obtain the IT3 FLS as

$$y_{mnIT3}(\mathbf{x}_k) = \frac{\sum_{\lambda=1}^{\rho} \left(\underline{\alpha}_\lambda y_{mn,\underline{\alpha}_\lambda}(\mathbf{x}_k) + \bar{\alpha}_\lambda y_{mn,\bar{\alpha}_\lambda}(\mathbf{x}_k)\right)}{\sum_{\lambda=1}^{\rho}(\underline{\alpha}_\lambda + \bar{\alpha}_\lambda)} \tag{6.44}$$

6. Optimize parameters $m$ and $n$ as in [12] for minimization of error.

$$\text{MSE} = \frac{1}{N}\sum_{k=1}^{N} \left(y_{mn,\text{IT3}}(\mathbf{x}_k) - y_k\right)^2 \tag{6.45}$$

### 6.4.1 Design Parameters

The nine control parameters in the proposed IT3 T-S fuzzy system are $c, m_1, m_2, \eta, \rho, w, \Delta, m, n$.

The number of fuzzy rules is indicated by the parameter $c$, which is typically preset by experience. The fuzziness in IT2 FCRM clustering is controlled by the fuzzifier parameters $m_1, m_2$. As suggested by Li et al. in [30], the parameter $\eta$ adjusts the fuzzy membership grades and has a value in the interval $[1, 100]$. The number of horizontal slices ($\alpha$-planes) is indicated by the parameter $\rho$. Depending on the application, a finer discretisation of $\alpha \in [0, 1]$ (large value of $\rho$) can result in higher accuracy. The triangular SMF's apex location is determined by the parameter $w \in [0, 1]$. Mendel et al. [7] presents the impact of this parameter. The uncertainty in secondary grades is determined by the constant parameter $\Delta \geq 1$. In order to improve application-related performance, the aforementioned parameters are typically adjusted through trial and error. The parameters $m, n$ can be adjusted during the design process and weight the sharing of each fired rule's lower and upper firing levels. Biglarbegian et al. in [12] propose the tuning algorithm for these $m, n$ parameters.

### 6.4.2 Computational Complexity Analysis

An IT2 FS has fewer design options than a GT2 FS, and an IT3 FS has even more than a GT2 FS. The number of control parameters and loop iterations of the IT2-FCRM clustering algorithms are the primary causes of the computation time in the suggested IT3 T-S model. The number of iterations $N_{itr}$, the number of $\alpha$-planes $\rho$, the number of data samples $N$, and the number of rules $c$ are the control parameters that contribute most. Calculation cost is described using the Basic Sequence Calculation (BSC), as described in [29]. For the upper and lower hyper-planes, this cost is $c \times N \times N_{itr}$ BSC. By using $\alpha$-plane representation with $\rho$ horizontal slices, the computational cost will be changed to $c \times N \times N_{itr} \times \rho$ BSC. Additionally, two levels of uncertainty in the $alpha$-plane, namely $[underlinealpha, baralpha]$, are introduced with the fuzziness parameter $Delta$. As a result, the computational cost is once more changed to $ctimesNtimesN_itrtimes2rho$ BSC. Other computational steps in a sequence calculation process are simple. Therefore, the computational cost for the proposed IT3 T-S FLS is roughly $c \times N \times N_{itr} \times 2\rho$ BSC.

Two-dimensional datasets with sizes of 250, 500, and 1000 are used to test the suggested algorithm. Using $c = 5$ and six $\alpha$-planes, the computational time for 50 iterations (IT2 FCRM clustering) is 3.64, 7.36, and 14.90 s, respectively. By changing $c$ and the number of $\alpha$-planes, additional computation time is also estimated. The computation time is found to increase roughly linearly with the size of the input data set $N$, the number of rules $c$, and the number of $\alpha$-planes ($\rho$).

## 6.5 Performance Evaluation

This section evaluates the IT3 T-S fuzzy model on the Box and Jenkins gas furnace benchmark. Noise, missing data, and invariant data are introduced to test its robustness, and the results are compared with recent fuzzy modeling methods.

### 6.5.1 Example (The Box-Jenkins System)

The proposed fuzzy model is demonstrated on the classical system identification problem introduced by Box and Jenkins. The process corresponds to a gas furnace, where the gas flow rate $u(k)$ serves as the input and the $CO_2$ concentration $y(k)$ as the output. The dataset includes 296 input–output pairs. Six input variables, $u(k)$, $u(k-1)$, $u(k-2)$, $y(k-1)$, $y(k-2)$, $y(k-3)$, are employed to predict $y(k)$. Of these, the first 148 samples are used for training and the remaining 148 for testing.

The fuzzy model parameters are specified as $c = 3$, $m_1 = 1.4$, $m_2 = 5$, $\eta = 51$, and $w = 0.5$. The $\alpha$-plane values are chosen as $[\alpha_1, \alpha_2, \alpha_3, \alpha_4, \alpha_5, \alpha_6] = [0, 0.2, 0.4, 0.6, 0.8, 1.0]$ with fuzziness $\Delta = 2$. The optimized parameters are

$m = 0.57$ and $n = 0.43$. Simulations are carried out with these settings for the cases described below.

Case A: Two training and two testing datasets (148 points each) are used. The first set contains clean data, while the second set includes Gaussian noise with a mean of 0 and a standard deviation of 0.2.

Case B: To model missing data, 30 samples of the first input $u(k)$ in both training and testing datasets (from Case A) are randomly set to zero. This is done with and without added noise.

Case C: To model invariability, 30 samples of the first input $u(k)$ in both training and testing datasets (from Case A) are replaced with their maximum values. This is also tested under noisy and noise-free conditions.

The prediction performance is shown in Fig. 6.3a and b.

### *6.5.2 Selection of Design Parameters*

In the simulation studies, the design parameters are chosen to achieve optimal accuracy. The number of fuzzy rules is set to closely match those used in comparable fuzzy modeling approaches. The $\alpha$-plane and fuzziness settings are predefined as $[\alpha_1, \alpha_2, \alpha_3, \alpha_4, \alpha_5, \alpha_6] = [0, 0.2, 0.4, 0.6, 0.8, 1.0]$ (equivalently $\rho = 6$) with $\Delta = 2$. The parameter $w \in [0, 1]$ is manually assigned values (0.5, 0.6, 0.8, 0.5, 0.5), following the discussion in [7]. The tuning parameters $m_1, m_2, \eta$ are varied across ranges ($m_1, m_2$ from 1.1 to 7.0 and $\eta$ from 1 to 100), and the corresponding mean square error (MSE) is recorded. The parameter set that yields the lowest MSE is selected. Initially, $m$ and $n$ are both set to 0.5. Once the other parameters are fixed, $m$ and $n$ are further refined using the update rules $m \leftarrow m + 0.01m$ and $n \leftarrow n - 0.01n$, according to the procedure described in [12].

## 6.6 Summary

A methodical approach to the design of an interval type-3 T-S fuzzy logic system based on theoretical base created for IT2 FLSs and GT2 FLSs has been presented in this chapter. The design methodology for IT3 FSs is developed using the $\alpha$-plane representation theorem that was proposed for GT2 FSs. Using hyper-plane-shaped membership functions and modified interval type-2 FCRM clustering algorithms, the baseline design for the IT2 T-S fuzzy system is implemented [29, 30]. The idea of interval type-2 secondary MF is used to elaborate the design process for IT3 T-S FLS. Therefore, the suggested method includes FOUs linked to both Primary MF (primary FOU) and Secondary MF (secondary FOU); this factor offers greater adaptability to handle a high level of uncertainty. By introducing noise into the simulated data and generating the uncertainties associated with missing and unvarying data, the effectiveness of the suggested approach is confirmed on a number of benchmark

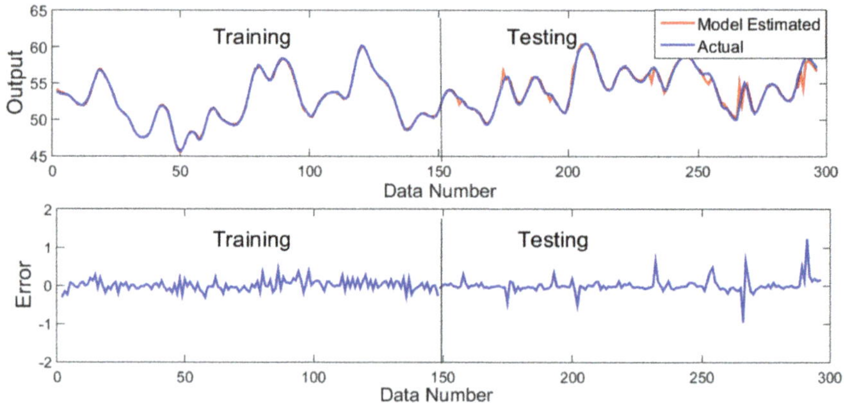


(a) Noise-free case

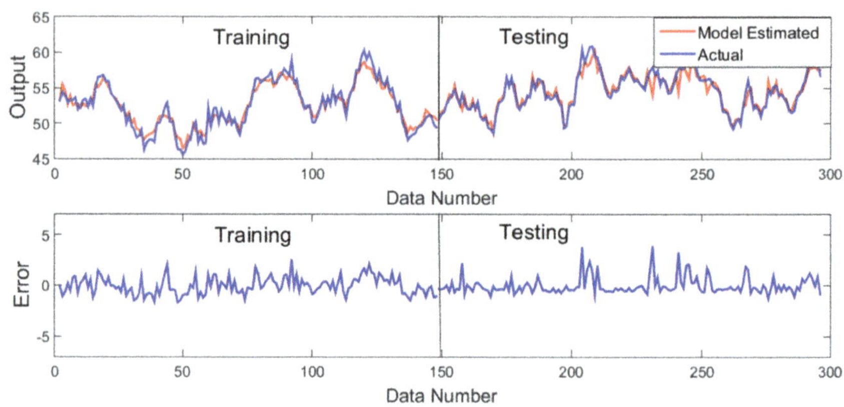


(b) Noisy case (0.2 STD)

**Fig. 6.3** Performance of the IT3 fuzzy model: **a** noise-free case, **b** noisy case (training and testing data sets with 0.2 STD noise level)

simulation studies. Additionally, the outcomes of the simulation are contrasted with the latest cutting-edge fuzzy modeling techniques.

# References


1. G.J Klir and M.J. Wierman, Uncertainty-Based Information, Heidelberg, Germany: Physica-Verlag, 1998
2. H.R Berenji, "Treatment of Uncertainty in Artificial Intelligence," in Machine Intelligence and Autonomy Aerospace systems, H. Heer and H.Lum, Eds., Washington, DC, USA: AIAA, 1988, pp. 233-247.
3. E Hisdal, "The IF THEN ELSE statement and interval valued fuzzy sets of higher type," International Journal of Man-Machine Studies, vol. 15, pp. 385-455, 1981, https://doi.org/10.1016/S0020-7373(81)80051-X.
4. J. M. Mendel, R. I. John and F. Liu, "Interval Type-2 Fuzzy Logic Systems Made Simple," in IEEE Transactions on Fuzzy Systems, vol. 14, no. 6, pp. 808-821, 2006, https://doi.org/10.1109/TFUZZ.2006.879986.
5. J. M. Mendel, "General Type-2 Fuzzy Logic Systems Made Simple: A Tutorial," in IEEE Transactions on Fuzzy Systems, vol. 22, no. 5, pp. 1162-1182, 2014, https://doi.org/10.1109/TFUZZ.2013.2286414.
6. F. Liu, "An efficient centroid type-reduction strategy for general type-2 fuzzy logic system," Information Sciences, vol. 178, pp. 2224-2236, 2008, https://doi.org/10.1016/j.ins.2007.11.014.
7. J. M. Mendel, F. Liu and D. Zhai, "$\alpha$-Plane Representation for Type-2 Fuzzy Sets: Theory and Applications," in IEEE Transactions on Fuzzy Systems, vol. 17, no. 5, pp. 1189-1207, 2009, https://doi.org/10.1109/TFUZZ.2009.2024411.
8. J. M. Mendel, "Comments on '$\alpha$-Plane Representation for Type-2 Fuzzy Sets: Theory and Applications'," in IEEE Transactions on Fuzzy Systems, vol. 18, no. 1, pp. 229-230, 2010, https://doi.org/10.1109/TFUZZ.2009.2039368.
9. J. M. Mendel, "Comparing the Performance Potentials of Interval and General Type-2 Rule-Based Fuzzy Systems in Terms of Sculpting the State Space," in IEEE Transactions on Fuzzy Systems, vol. 27, no. 1, pp. 58-71, 2019, https://doi.org/10.1109/TFUZZ.2018.2856184.
10. S. Kulkarni, R. Agrawal and F. Chung-Hoon Rhee, "Determining the Optimal Fuzzifier Range for Alpha-Planes of General Type-2 Fuzzy Sets," 2018 IEEE International Conference on Fuzzy Systems (FUZZ-IEEE), Rio de Janeiro, 2018, pp. 1-8, https://doi.org/10.1109/FUZZ-IEEE.2018.8491556.
11. E. Ontiveros, P. Melin and O. Castillo, "High order alpha-planes integration: A new approach to computational cost reduction of general type-2 fuzzy systems," Engineering Applications of Artificial Intelligence, vol. 74, pp. 186-197, 2018, https://doi.org/10.1016/j.engappai.2018.06.013.
12. M. Biglarbegian, W. W. Melek and J. M. Mendel, "On the Stability of Interval Type-2 TSK Fuzzy Logic Control Systems," in IEEE Transactions on Systems, Man, and Cybernetics, Part B (Cybernetics), vol. 40, no. 3, pp. 798-818, 2010, https://doi.org/10.1109/TSMCB.2009.2029986.
13. J. M. Mendel, Uncertain Rule-Based Fuzzy Logic Systems: Introduction and New Directions, NJ, Upper Saddle River: Prentice-Hall, 2001.
14. J. M. Mendel, "Plotting 2-1/2 D figures for general type-2 fuzzy sets by hand or by PowerPoint," 2012 IEEE International Conference on Fuzzy Systems, Brisbane, QLD, 2012, pp. 1-8, https://doi.org/10.1109/FUZZ-IEEE.2012.6251270.
15. D. Wu and J. M. Mendel, "Similarity Measures for Closed General Type-2 Fuzzy Sets: Overview, Comparisons, and a Geometric Approach," in IEEE Transactions on Fuzzy Systems, vol. 27, no. 3, pp. 515-526, 2019, https://doi.org/10.1109/TFUZZ.2018.2862869.
16. A. Sakalli, T. Kumbasar and J. M. Mendel, "Towards Systematic Design of General Type-2 Fuzzy Logic Controllers: Analysis, Interpretation and Tuning," in IEEE Transactions on Fuzzy Systems, early access, https://doi.org/10.1109/TFUZZ.2020.3016034.
17. J. Z. Shi, "A Fractional Order General Type-2 Fuzzy PID Controller Design Algorithm," in IEEE Access, vol. 8, pp. 52151-52172, 2020, https://doi.org/10.1109/ACCESS.2020.2980686.

18. F. Baghbani, M.-R. Akbarzadeh-T and A. Akbarzadeh, "Indirect adaptive robust mixed H2 / H∞ general type-2 fuzzy control of uncertain nonlinear systems," Applied Soft Computing, vol. 72, pp. 392-418, 2018, https://doi.org/10.1016/j.asoc.2018.06.049.
19. J. Andreu-Perez, F. Cao, H. Hagras and G. Yang, "A Self-Adaptive Online Brain–Machine Interface of a Humanoid Robot Through a General Type-2 Fuzzy Inference System," in IEEE Transactions on Fuzzy Systems, vol. 26, no. 1, pp. 101-116, 2018, https://doi.org/10.1109/TFUZZ.2016.2637403.
20. L. A. Torres Salomao, M. Mahfouf, E. El-Samahy and C. Ting, "Psychophysiologically Based Real-Time Adaptive General Type 2 Fuzzy Modeling and Self-Organizing Control of Operator's Performance Undertaking a Cognitive Task," in IEEE Transactions on Fuzzy Systems, vol. 25, no. 1, pp. 43-57, Feb. 2017, https://doi.org/10.1109/TFUZZ.2016.2598363.
21. H. Hassani, J. Zarei, M. M. Arefi and R. Razavi-Far, "zSlices-Based General Type-2 Fuzzy Fusion of Support Vector Machines With Application to Bearing Fault Detection," in IEEE Transactions on Industrial Electronics, vol. 64, no. 9, pp. 7210-7217, 2017, https://doi.org/10.1109/TIE.2017.2688963.
22. A. Mohammadzadeh, M. H. Sabzalian and W. Zhang, "An Interval Type-3 Fuzzy System and a New Online Fractional-Order Learning Algorithm: Theory and Practice," in IEEE Transactions on Fuzzy Systems, vol. 28, no. 9, pp. 1940-1950, 2020, https://doi.org/10.1109/TFUZZ.2019.2928509.
23. J. T. Rickard, J. Aisbett and G. Gibbon, "Fuzzy Subsethood for Fuzzy Sets of Type-2 and Generalized Type- n," in IEEE Transactions on Fuzzy Systems, vol. 17, no. 1, pp. 50-60, 2009, https://doi.org/10.1109/TFUZZ.2008.2006369.
24. M. F. Zarandi, R. Gamasaee and O. Castillo, "Type-1 to type-n fuzzy logic and systems," in Fuzzy Logic in Its 50th Year, C. Kahraman et al. , Eds., Switzerland: Springer, 2016, pp. 129-157, https://doi.org/10.1007/978-3-319-31093-0_6.
25. O. Castillo, "Towards finding the optimal n in designing type-n fuzzy systems for particular classes of problems: A review," Applied and Computational Mathematics, vol. 17, no. 1, pp. 3-9, 2018.
26. B. Zhu, Z. Xu and M. Xia, "Dual hesitant fuzzy sets," Journal of Applied Mathematics, vol. 2012, 2012, https://doi.org/10.1155/2012/879629.
27. V. Torra and Y. Narukawa, "On hesitant fuzzy sets and decision," 2009 IEEE International Conference on Fuzzy Systems, Jeju Island, 2009, pp. 1378-1382, https://doi.org/10.1109/FUZZY.2009.5276884.
28. I. B. Türkşen, "Type 2 representation and reasoning for CWW," Fuzzy Sets and Systems, vol. 127, no. 1, pp. 17-36, 2002, https://doi.org/10.1016/S0165-0114(01)00150-6.
29. W. Zou, C. Li and N. Zhang, "A T–S Fuzzy Model Identification Approach Based on a Modified Inter Type-2 FRCM Algorithm," in IEEE Transactions on Fuzzy Systems, vol. 26, no. 3, pp. 1104-1113, 2018, https://doi.org/10.1109/TFUZZ.2017.2704542.
30. C. Li, J. Zhou, L. Chang, Z. Huang and Y. Zhang, "T–S Fuzzy Model Identification Based on a Novel Hyperplane-Shaped Membership Function," in IEEE Transactions on Fuzzy Systems, vol. 25, no. 5, pp. 1364-1370, 2017, https://doi.org/10.1109/TFUZZ.2016.2598850.
31. M. H. Fazel Zarandi, R. Gamasaee and I. B. Turksen, "A type-2 fuzzy c-regression clustering algorithm for Takagi–Sugeno system identification and its application in the steel industry," Information Sciences, vol. 187, no. 15, pp. 179-203, Mar. 2012, https://doi.org/10.1016/j.ins.2011.10.015.
32. C. Li, J. Zhou, X. Xiang, Q. Li and X. An, "T–S fuzzy model identification based on a novel fuzzy c-regression model clustering algorithm," Engineering Applications of Artificial Intelligence, vol. 22, no. 4-5, pp. 646-653, 2009, https://doi.org/10.1016/j.engappai.2009.02.003.
33. C. Li, J. Zhou, B. Fu, P. Kou and J. Xiao, "T–S Fuzzy Model Identification With a Gravitational Search-Based Hyperplane Clustering Algorithm," in IEEE Transactions on Fuzzy Systems, vol. 20, no. 2, pp. 305-317, 2012, https://doi.org/10.1109/TFUZZ.2011.2173693.
34. Euntai Kim, Minkee Park, Seungwoo Kim and Mignon Park, "A transformed input-domain approach to fuzzy modeling," in IEEE Transactions on Fuzzy Systems, vol. 6, no. 4, pp. 596-604, 1998, https://doi.org/10.1109/91.728458.

35. C. Juang, R. Huang and W. Cheng, "An Interval Type-2 Fuzzy-Neural Network With Support-Vector Regression for Noisy Regression Problems," in IEEE Transactions on Fuzzy Systems, vol. 18, no. 4, pp. 686-699, 2010, https://doi.org/10.1109/TFUZZ.2010.2046904.
36. C. Juang and Y. Tsao, "A Self-Evolving Interval Type-2 Fuzzy Neural Network With Online Structure and Parameter Learning," in IEEE Transactions on Fuzzy Systems, vol. 16, no. 6, pp. 1411-1424, 2008, https://doi.org/10.1109/TFUZZ.2008.925907.
37. S. W. Tung, C. Quek and C. Guan, "eT2FIS: An evolving type-2 neural fuzzy inference system," Information Sciences, vol. 220, pp. 124-148, 2013, https://doi.org/10.1016/j.ins.2012.02.031.
38. N. J. Cheung, X. Ding and H. Shen, "OptiFel: A Convergent Heterogeneous Particle Swarm Optimization Algorithm for Takagi–Sugeno Fuzzy Modeling," in IEEE Transactions on Fuzzy Systems, vol. 22, no. 4, pp. 919-933, 2014, https://doi.org/10.1109/TFUZZ.2013.2278972.
39. A. Bagis, "Fuzzy rule base design using tabu search algorithm for nonlinear system modeling," ISA Transactions, vol. 47, no.1, pp.32-44, 2008, https://doi.org/10.1016/j.isatra.2007.09.001.
40. M. Sugeno and T. Yasukawa, "A fuzzy-logic-based approach to qualitative modeling," in IEEE Transactions on Fuzzy Systems, vol. 1, no. 1, pp. 7-, Feb. 1993, https://doi.org/10.1109/TFUZZ.1993.390281.
41. J Abony and B Feil, Cluster analysis for data mining and system identification, Switzerland: Birkhauser Verlag AG, 2007.
42. Chia-Feng Juang and Chin-Teng Lin, "An online self-constructing neural fuzzy inference network and its applications," in IEEE Transactions on Fuzzy Systems, vol. 6, no. 1, pp. 12-32, 1998, https://doi.org/10.1109/91.660805.
43. Jung-Hsien Chiang and Pei-Yi Hao, "Support vector learning mechanism for fuzzy rule-based modeling: a new approach," in IEEE Transactions on Fuzzy Systems, vol. 12, no. 1, pp. 1-12, 2004, https://doi.org/10.1109/TFUZZ.2003.817839.
44. J. Aisbett, J. T. Rickard and D. Morgenthaler, "Intersection and union of type-n fuzzy sets," International Conference on Fuzzy Systems, 2010, pp. 1-8, https://doi.org/10.1109/FUZZY.2010.5584174.
45. J. T. Rickard, J. Aisbett, G. Gibbon and D. Morgenthaler, "Fuzzy subsethood for Type-n fuzzy sets," NAFIPS 2008 - 2008 Annual Meeting of the North American Fuzzy Information Processing Society, 2008, pp. 1-6, https://doi.org/10.1109/NAFIPS.2008.4531276.
46. P. Fisher, T. Cheng and J. Wood, "Higher Order Vagueness in Geographical Information: Empirical Geographical Population of Type n Fuzzy Sets," Geoinformatica, vol. 11, no. 3, pp. 311–330, 2007, https://doi.org/10.1007/s10707-006-0009-5.
47. I. B. Türkşen, "From Type 1 to Full Type N Fuzzy System Models," Journal of Multiple-Valued Logic and Soft Computing, vol. 22, no. 4-6, pp. 543-560, 2014.
48. A. Mohammadzadeh, O. Castillo, S. S Band and A. Mosavi, "A Novel Fractional-Order Multiple-Model Type-3 Fuzzy Control for Nonlinear Systems with Unmodeled Dynamics," International Journal of Fuzzy Systems, early access ,2021, https://doi.org/10.1007/s40815-021-01058-1.
49. Z. Liu, A. Mohammadzadeh, H. Turabieh, M. Mafarja, S. S. Band and A. Mosavi, "A New Online Learned Interval Type-3 Fuzzy Control System for Solar Energy Management Systems," in IEEE Access, vol. 9, pp. 10498-10508, 2021, https://doi.org/10.1109/ACCESS.2021.3049301.
50. C. Ma, A. Mohammadzadeh, H. Turabieh, M. Mafarja, S. S. Band and A. Mosavi, "Optimal Type-3 Fuzzy System for Solving Singular Multi-Pantograph Equations," in IEEE Access, vol. 8, pp. 225692-225702, 2020, https://doi.org/10.1109/ACCESS.2020.3044548.
51. W. Pedrycz, "Fuzzy Sets of Higher Type and Higher Order in Fuzzy Modeling," in Frontiers of Higher Order Fuzzy Sets, A. Sadeghian and H. Tahayori, Eds., New York, NY, USA: Springer, 2015, pp. 31-49, https://doi.org/10.1007/978-1-4614-3442-9_3.
52. K. Mittal, A. Jain, K. S Vaisla, O. Castillo and J. Kacprzyk, "A comprehensive review on type 2 fuzzy logic applications: Past, present and future," Engineering Applications of Artificial Intelligence, vol. 95, 2020, https://doi.org/10.1016/j.engappai.2020.103916.
53. E. Ontiveros-Robles, P. Melin, O. Castillo, "Comparative analysis of noise robustness of type 2 fuzzy logic controllers," Kybernetika, vol. 54, no. 1, pp. 175-201, 2018.

54. E. Ontiveros, P. Melin, O. Castillo, "Comparative study of interval Type-2 and general Type-2 fuzzy systems in medical diagnosis," Information Sciences, vol. 525, pp. 37-53, 2020, https://doi.org/10.1016/j.ins.2020.03.059.
55. P. D'Alterio, J. M. Garibaldi, R. I. John and A. Pourabdollah, "Constrained Interval Type-2 Fuzzy Sets," in IEEE Transactions on Fuzzy Systems, vol. 29, no. 5, pp. 1212-1225, 2021, https://doi.org/10.1109/TFUZZ.2020.2970911.
56. R. Boukezzoula, L. Jaulin, B. Desrochers and D. Coquin, "Thick Fuzzy Sets (TFSs) and Their Potential Use in Uncertain Fuzzy Computations and Modeling," in IEEE Transactions on Fuzzy Systems, early access, https://doi.org/10.1109/TFUZZ.2020.3018550.
57. C. Chen, D. Wu, J. M. Garibaldi, R. I. John, J. Twycross and J. M. Mendel, "A Comprehensive Study of the Efficiency of Type-Reduction Algorithms," in IEEE Transactions on Fuzzy Systems, early access, https://doi.org/10.1109/TFUZZ.2020.2981002.
58. A. S. Weigend and N. A. Gersehnfield, Time Series Prediction: Forecasting the Future and Understanding the Past. Reading, MA: Addison-Wesley, 1994.
59. S. Singh, "Noise impact on time-series forecasting using an intelligent pattern matching technique," Pattern Recognition, vol. 32, no. 8, pp. 1389-1398, 1999, https://doi.org/10.1016/S0031-3203(98)00174-5.
60. J. M. Mendel, "Computing derivatives in interval type-2 fuzzy logic systems," in IEEE Transactions on Fuzzy Systems, vol. 12, no. 1, pp. 84-98, 2004, https://doi.org/10.1109/TFUZZ.2003.822681.
61. George E. Tsekouras, "On the use of the weighted fuzzy c-means in fuzzy modeling," Advances in Engineering Software, vol. 36, no. 5, pp. 287-300, 2005, https://doi.org/10.1016/j.advengsoft.2004.12.001.
62. M. Luo, F. Sun and H. Liu, "Hierarchical Structured Sparse Representation for T–S Fuzzy Systems Identification," in IEEE Transactions on Fuzzy Systems, vol. 21, no. 6, pp. 1032-1043, 2013, https://doi.org/10.1109/TFUZZ.2013.2240690.
63. I. Eyoh, R. John, G. De Maere and E. Kayacan, "Hybrid Learning for Interval Type-2 Intuitionistic Fuzzy Logic Systems as Applied to Identification and Prediction Problems," in IEEE Transactions on Fuzzy Systems, vol. 26, no. 5, pp. 2672-2685, 2018, https://doi.org/10.1109/TFUZZ.2018.2803751.
64. I.J. Eyoh, U.A. Umoh, U.G. Inyang and J.E. Eyoh, "Derivative-Based Learning of Interval Type-2 Intuitionistic Fuzzy Logic Systems for Noisy Regression Problems," International Journal of Fuzzy Systems, vol. 22, no. 3, pp. 1007–1019, 2020, https://doi.org/10.1007/s40815-020-00806-z.

# Chapter 7
# Interval Type-3 T-S Fuzzy System for Nonlinear Aerodynamic Modeling

**Abstract** This chapter presents an Interval Type-3 (IT3) T-S fuzzy system for nonlinear aerodynamic modeling using recorded flight data from the ATTAS aircraft under quasi-steady stall conditions. The model estimates longitudinal aerodynamic coefficients and accounts for lift, drag, and pitching moment. To evaluate robustness, external noise and uncertainties from missing or invariant data are introduced. Results show that the IT3 fuzzy approach effectively captures complex aircraft dynamics under uncertain conditions and outperforms existing fuzzy models, particularly as noise and uncertainty increase. The study highlights the strong predictive ability and industrial relevance of the IT3 fuzzy system.

## 7.1 Introduction

Aerodynamic modeling and system identification processes are widely employed to visualize and examine the intricate nonlinear characteristics of aircraft, spanning from the initial stages of design through improvements, upgrades, fault detection, and the creation of flight control laws. Numerous system identification techniques have been documented in the literature and have been effectively utilized for aerodynamic modeling and estimating aircraft parameters [1–4]. AI-based modeling methods [5] (neural networks and fuzzy systems) are able to predict the aircraft dynamics even under high degree of cross-coupling and data uncertainties that arise during close combat flying maneuvers [6–9]. Aircraft dynamics in such movements are difficult to describe by physical laws, and modeling their internal mechanics is growing more challenging [10].

These nonlinear complex aerodynamics can be modeled using AI-based fuzzy logic theory, which has certain unique benefits over traditional mathematical modeling. These benefits include: (1) a methodology for human-understandable reasoning; (2) the capacity to integrate linguistic and numerical data; (3) the ability to function effectively in the presence of random disturbances and noisy measurements; and (4) the ability to represent non-linear complex functions using local linear simplified models [11, 12].

 
D. J. Singh et al., *Fuzzy Systems with Applications to Modeling and Control of an Aircraft*, Algorithms for Intelligent Systems, https://doi.org/10.1007/978-981-95-7888-7_7

Type-1 (T1) and Type-2 (T2) fuzzy logic theories are created and successfully implemented in a variety of application areas in the design of fuzzy logic systems. Fuzzy Logic Systems (FLSs) of T2 are better at handling high levels of noise and data uncertainty. The Interval Type-2 (IT2) FLS is less complex than the General Type-2 (GT2) FLS [13–15].

A comprehensive review of neural network-based aerodynamic modeling is reported in [16–20], while methods for unsteady aerodynamics are discussed in [21–23]. Deep learning techniques for nonlinear aerodynamic modeling are presented in [24, 25].

Recent studies have explored interval type-3 fuzzy systems (FSs) as an extension of GT2 FSs. Theoretical advances in fuzzy subset-hood for type-$n$ FSs are presented in [26], while type-3 FSs are outlined as future research directions in [27, 28]. Initial theoretical foundations of IT3 FLS are given by Mohammadzadeh et al. [29], and a novel IT3-based control method for uncertain nonlinear systems is proposed in [30]. Building on [29], an IT3 fuzzy controller for solar energy management is developed in [31].

Singh et al. [32] proposed a novel method for designing IT3 T-S FSs. Their approach employs the $\alpha$-plane theorem [33, 34] for IT3 FLS design, along with hyperplane-shaped membership functions and modified IT2 FCRM clustering algorithms [35, 36].

Building on the advantages of IT3 FLSs, this work applies the framework to nonlinear aircraft dynamics modeling using recorded flight test data. The dataset, obtained from the Advanced Technology Testing Aircraft System (ATTAS) under quasi-steady stall maneuvers, serves as the basis for developing the IT3 T-S fuzzy longitudinal aerodynamic model [32]. The modeling focuses on the pitching moment, lift, and drag coefficients. To assess robustness, noise and uncertainties from missing and invariant data are introduced into the flight data. The performance of the proposed IT3 T-S fuzzy models is then compared against recent system identification techniques.

## 7.2 Longitudinal Aircraft Dynamics

The longitudinal dynamics of an aircraft are nonlinear functions of state variables such as flight speed $V$, angle of attack $\alpha$, pitch angle $\theta$, and pitch rate $q$. The corresponding aerodynamic forces and moments acting on the aircraft are governed by the dynamic equations. The nonlinear longitudinal dynamics can be expressed mathematically as shown in [37, 38]:

$$m\dot{V} = -D - mg\sin\gamma + T\cos\alpha, \tag{7.1}$$

$$mV\dot{\alpha} = -L + mg\cos\gamma - T\sin\alpha + mVq, \tag{7.2}$$

$$\dot{\theta} = q, \tag{7.3}$$

$$I_y \dot{q} = M, \tag{7.4}$$

Here, $m$ denotes the aircraft mass, $g$ the gravitational acceleration, $I_y$ the moment of inertia about the $y$-axis, and $y = \theta - \alpha$ the flight path angle. The aerodynamic forces drag $(D)$, lift $(L)$, and the pitching moment $(M)$ are expressed as nonlinear functions of the control input and state variables, which can be represented as:

$$L = \tfrac{1}{2}\rho V^2 S C_L(\alpha, \delta_{el}, \dot{\alpha}, q), \tag{7.5}$$

$$D = \tfrac{1}{2}\rho V^2 S C_D(\alpha, \delta_{el}), \tag{7.6}$$

$$M = \tfrac{1}{2}\rho V^2 S\bar{c} C_M(\alpha, \delta_{el}, \dot{\alpha}, q), \tag{7.7}$$

where $S, \rho, \bar{c}$ are wing surface area, air density, and mean chord length. The input to the longitudinal model is elevator deflection $(\delta_{el})$. The parameters $C_D$, $C_L$, and $C_M$ are aerodynamic forces and moment derivatives, respectively. Equations (7.5), (7.6), and (7.7) are generally used to characterize the longitudinal aerodynamic model of the fixed-wing aircraft. The flow field over the aircraft body is laminar and remains linked to it when flying at low or moderate angles of attack. The desired lift is produced by this connected flow over the wing. When the pitch rate, angle-of-attack, and elevator deflection parameters are altered, the flow field's shape also changes. High-angle-of-attack unstable aerodynamics is more difficult to model and forecast than low or moderate angle-of-attack flying circumstances. High-angle-of-attack aerodynamics is defined as a three-dimensional, time-varying, nonlinear phenomenon in which the aircraft's forces and moments are significantly impacted by separated airflow. There is a significant loss of lift as a result of flow separation [10, 37]. The following nonlinear function describes the aerodynamic force and moment coefficients under such flight conditions.

$$C_D,\ C_L,\ C_M = f_{\text{nonlinear}}(\alpha, q, \delta_e) \tag{7.8}$$

In this chapter, longitudinal aerodynamic coefficients $(C_D, C_L, C_M)$ are modeled utilizing the type-3 fuzzy logic approach of the interval from the recorded flight data.

## 7.3 Architecture of IT3 T-S Fuzzy Longitudinal Aerodynamic Model

Nonlinear aerodynamic modeling is carried out using an IT3 T-S fuzzy system with flight data from the VFW-614 ATTAS aircraft under quasi-steady stall conditions [4, 39]. The input parameters are angle of attack $(\alpha)$, pitch rate $(q)$, and elevator deflection $(\delta_e)$, while the outputs are aerodynamic coefficients $(C_D, C_L, C_M)$. These outputs are derived from measured accelerations $(a_x, a_y, a_z)$ and angular rates

$(p, q, r)$ through a preprocessing step [4]. The related expressions for data preprocessing are [4] as:

- A 15-point digital filter is used to reduce high-frequency noise in the measured raw data
- Linear accelerations measured across three axes are mapped to the aircraft's center of gravity.
- Angular rates are differentiated using numerical schemes and smoothed to obtain angular accelerations $(\dot{p}, \dot{q}, \dot{r})$.
- Aerodynamic force coefficients are derived from aircraft and engine parameters.
- Lift and drag coefficients are resolved from body-axis forces, while pitching moment coefficients are obtained using inertia, geometry, and engine data, then mapped to the aerodynamic center.

The methodology for estimating lift, drag, and pitching moment within the IT3 T-S fuzzy framework is detailed in Sects. 7.3.1, 7.3.2, and 7.3.3, providing an aerodynamic-oriented design approach.

### 7.3.1 IT3 T-S Fuzzy Model for Drag Force Coefficient

As discussed in [32], the IT2 and GT2 FLS frameworks serve as the foundation for constructing the IT3 T-S fuzzy system. For the basic IT2 T-S FLS design, the Hyper-Plane-Shaped Membership Function (HPSMF) and the modified type-2 Fuzzy C-Regression Model (FCRM) clustering algorithm are employed [35, 36]. Several mathematical formulations from [32] are adapted to suit the requirements of drag coefficient modeling.

Assume that data pairs $\left([\alpha_k\ q_k\ \delta_{e_k}], C_{D_k}\right)$, are organized in $c$ clusters for $k = 1, 2, \ldots, N$ and $i$th cluster is in form of hyperplane regression function as:

$$\hat{C}_{D_k} = f^i\left([\alpha_k\ q_k\ \delta_{e_k}], \boldsymbol{\omega}_{D_i}\right) = a^i_{D\alpha}\alpha_k + a^i_{Dq}q_k + a^i_{D\delta_e}\delta_{e_k} + a^i_{D0}, \tag{7.9}$$

where, $[\alpha_k\ q_k\ \delta_{ek}]$, is the $k$th input vector, $\hat{C}_{D_k}$ is $k$th estimated output vector and the number of clusters are $i = 1, 2, \cdots, c$. The hyperplane model coefficient vector is $\boldsymbol{\omega}_{D_i} = \left[a^i_{D\alpha},\ a^i_{Dq},\ a^i_{D\delta_e},\ a^i_{D0}\right]$.

To formulate the IT2 FS, fuzzifiers are considered for IT2-FCRM modeling [32].

The data sets are segregated into hyperplane-shaped clusters by using the above algorithms, and the updated membership function matrices are obtained as $\overline{\mathbf{U}}_D = [\overline{u}_{Dik}]_{c\times N}$, $\underline{\mathbf{U}}_D = [\underline{u}_{Dik}]_{c\times N}$.

The Weighted Least Squares (WLS) method is used to obtain upper and lower hyperplanes as [13, 40]

$$\overline{\boldsymbol{\omega}}_{D_i} = \left[\mathbf{X}^T\overline{\mathbf{P}}_{D_i}\mathbf{X}\right]^{-1}\mathbf{X}^T\overline{\mathbf{P}}_{D_i}\mathbf{C}_D, \tag{7.10}$$

$$\underline{\boldsymbol{\omega}}_{D_i} = \left[\mathbf{X}^T \underline{\mathbf{P}}_{D_i} \mathbf{X}\right]^{-1} \mathbf{X}^T \underline{\mathbf{P}}_{D_i} \mathbf{C}_D, \tag{7.11}$$

where, $\mathbf{X} = [\alpha_k \;\; q_k \;\; \delta_{e_k} \;\; 1]_{N\times 4}$ and $C_D = [C_{Dk}]_{N\times 1}$. The matrices $\overline{\mathbf{P}}_{D_i}$ and $\underline{\mathbf{P}}_{D_i}$ are as follows:

$$\overline{\mathbf{P}}_{D_i} = \begin{bmatrix} \overline{u}_{Di1} & 0 & \cdots & 0 \\ 0 & \overline{u}_{Di2} & \cdots & 0 \\ \vdots & \vdots & \ddots & \vdots \\ 0 & 0 & \cdots & \overline{u}_{DiN} \end{bmatrix}, \quad \underline{\mathbf{P}}_{D_i} = \begin{bmatrix} \underline{u}_{Di1} & 0 & \cdots & 0 \\ 0 & \underline{u}_{Di2} & \cdots & 0 \\ \vdots & \vdots & \ddots & \vdots \\ 0 & 0 & \cdots & \underline{u}_{DiN} \end{bmatrix} \in R^{N\times N}.$$

The hyperplane-shaped Gaussian MF projected in [36] is utilized for each input parameter. The upper and lower membership degrees for each input parameter are estimated from a hyperplane-shaped Gaussian MF as:

For first input parameter $\alpha_k$:

$$\overline{\mu}_{\tilde{D}_i}(\alpha_k) = \exp\left(-\eta_D \cdot \frac{d_{Di\alpha_k}(\overline{\boldsymbol{\omega}}_{D_i})}{\max\left(\{d_{Di\alpha_k}(\overline{\boldsymbol{\omega}}_{D_i})\},\; i = 1, \ldots, c\right)}\right), \tag{7.12a}$$

$$\underline{\mu}_{\tilde{D}_i}(\alpha_k) = \exp\left(-\eta_D \cdot \frac{d_{Di_{\alpha_k}}(\underline{\boldsymbol{\omega}}_{Di})}{\max(\{d_{Di_{\alpha_k}}(\underline{\boldsymbol{\omega}}_{Di})\}, i = 1, \ldots, c)}\right), \tag{7.12b}$$

where, $d_{Di\alpha_k}(\bar{\boldsymbol{\omega}}_{Di})$ represent the distance of $\alpha_k$ from $i$th upper and $d_{Di\alpha_k}(\underline{\boldsymbol{\omega}}_{Di})$ is the distance of $\alpha_k$ from $i$th lower hyper planes. The parameter $\eta_D$ is a constant, which is used for tuning of upper and lower membership grades $\bar{\mu}_{\tilde{D}_i}(\alpha_k)$ and $\underline{\mu}_{\tilde{D}_i}(\alpha_k)$. These parameters regulate the fuzziness of the membership grades and are referred to as the antecedent membership function parameters. The parameters $d_{Di\alpha_k}(\bar{\boldsymbol{\omega}}_{Di})$ and $d_{Di\alpha_k}(\underline{\boldsymbol{\omega}}_{Di})$ are obtained as follows:

$$d_{Di\alpha_k}(\bar{\boldsymbol{\omega}}_{Di}) = \frac{|\alpha_k \cdot \bar{\boldsymbol{\omega}}_{Di}|}{\|\bar{\boldsymbol{\omega}}_{Di}\|} \quad \text{and} \quad d_{Di\alpha_k}(\underline{\boldsymbol{\omega}}_{Di}) = \frac{|\alpha_k \cdot \underline{\boldsymbol{\omega}}_{Di}|}{\|\underline{\boldsymbol{\omega}}_{Di}\|}$$

For second input parameter $q_k$:

$$\bar{\mu}_{\tilde{D}_i}(q_k) = \exp\left(-\eta_D \cdot \frac{d_{Diq_k}(\bar{\boldsymbol{\omega}}_{Di})}{\max(\{d_{Diq_k}(\bar{\boldsymbol{\omega}}_{Di})\}, i = 1, \ldots, c)}\right), \tag{7.13a}$$

$$\underline{\mu}_{\tilde{D}_i}(q_k) = \exp\left(-\eta_D \cdot \frac{d_{Diq_k}(\underline{\boldsymbol{\omega}}_{Di})}{\max(\{d_{Diq_k}(\underline{\boldsymbol{\omega}}_{Di})\}, i = 1, \ldots, c)}\right), \tag{7.13b}$$

where, $d_{Diq_k}(\bar{\omega}_{Di})$ is the distance of $q_k$ from $i$th upper and $d_{Diq_k}(\underline{\omega}_{Di})$ is the distance of $q_k$ from $i$th lower hyper planes

$$d_{Diq_k}(\bar{\boldsymbol{\omega}}_{Di}) = \frac{|q_k \cdot \bar{\boldsymbol{\omega}}_{Di}|}{\|\bar{\boldsymbol{\omega}}_{Di}\|} \quad \text{and} \quad d_{Diq_k}(\underline{\boldsymbol{\omega}}_{Di}) = \frac{|q_k \cdot \underline{\boldsymbol{\omega}}_{Di}|}{\|\underline{\boldsymbol{\omega}}_{Di}\|}$$

For third input parameter $\delta_{ek}$:

$$\bar{\mu}_{\tilde{D}_i}(\delta_{ek}) = \exp\left(-\eta_D \cdot \frac{d_{Di\delta_{ek}}(\bar{\boldsymbol{\omega}}_{Di})}{\max(\{d_{Di\delta_{ek}}(\bar{\boldsymbol{\omega}}_{Di}), i = 1, \ldots, c\})}\right), \tag{7.14a}$$

$$\underline{\mu}_{\tilde{D}_i}(\delta_{ek}) = \exp\left(-\eta_D \cdot \frac{d_{Di\delta_{ek}}(\underline{\boldsymbol{\omega}}_{Di})}{\max(\{d_{Di\delta_{ek}}(\underline{\boldsymbol{\omega}}_{Di}), i = 1, \ldots, c\})}\right), \tag{7.14b}$$

where, $d_{Di\delta_{ek}}(\bar{\omega}_{Di})$ is the distance of $\delta_{ek}$ from $i$th upper and $d_{Di\delta_{ek}}(\underline{\omega}_{Di})$ is the distance of $\delta_{ek}$ from $i$th lower hyper planes

$$d_{Di\delta_{ek}}(\bar{\boldsymbol{\omega}}_{Di}) = \frac{|\delta_{ek} \cdot \bar{\boldsymbol{\omega}}_{Di}|}{\|\bar{\boldsymbol{\omega}}_{Di}\|} \quad \text{and} \quad d_{Di\delta_{ek}}(\underline{\boldsymbol{\omega}}_{Di}) = \frac{|\delta_{ek} \cdot \underline{\boldsymbol{\omega}}_{Di}|}{\|\underline{\boldsymbol{\omega}}_{Di}\|}$$

An IT2 T-S FLS for modeling of the Drag coefficient is expressed by the following IF-THEN fuzzy rules ( $i$th rule)

$$R_D^i : \text{ IF } \alpha \text{ is } \tilde{D}_\alpha^i \text{ and } q \text{ is } \tilde{D}_q^i \text{ and } \delta_e \text{ is } \tilde{D}_{\delta_e}^i$$

THEN

$$C_D^i(\alpha, q, \delta_e) = C_{D\alpha}^i\, \alpha + C_{Dq}^i\, q + C_{D\delta_e}^i\, \delta_e + C_{D0}^i,$$

where $\tilde{D}_\alpha^i$, $\tilde{D}_q^i$, $\tilde{D}_{\delta_e}^i$ are fuzzy model terms for the drag coefficient model related to each input parameter.

The interval type consequent parameters $\left(C_{D\alpha}^i, C_{Dq}^i, C_{D\delta_e}^i, C_{D0}^i\right)$ of IT2 T-S FLS are expressed as [40]:

$$C_{D\alpha}^i = \left[\nu_{D\alpha}^i - s_{D\alpha}^i,\ \nu_{D\alpha}^i + s_{D\alpha}^i\right],$$

$$C_{Dq}^i = \left[\nu_{Dq}^i - s_{Dq}^i,\ \nu_{Dq}^i + s_{Dq}^i\right],$$

$$C_{D\delta_e}^i = \left[\nu_{D\delta_e}^i - s_{D\delta_e}^i,\ \nu_{D\delta_e}^i + s_{D\delta_e}^i\right],$$

$$C_{D0}^i = \left[\nu_{D0}^i - s_{D0}^i,\ \nu_{D0}^i + s_{D0}^i\right],$$

where, $\left(v^i_{D\alpha}, v^i_{Dq}, v^i_{D\delta_e}, v^i_{D0}\right)$ signifies the mean (center) of $\left(C^i_{D\alpha}, C^i_{Dq}, C^i_{D\delta_e}, C^i_{D0}\right)$, and $\left(s^i_{D\alpha}, s^i_{Dq}, s^i_{D\delta_e}, s^i_{D0}\right)$ signifies the spread of $\left(C^i_{D\alpha}, C^i_{Dq}, C^i_{D\delta_e}, C^i_{D0}\right)$. These parameters are estimated from the coefficients of the upper and lower hyperplanes $\bar{\omega}_{Di} = \left(\bar{a}^i_{D\alpha}, \bar{a}^i_{Dq}, \bar{a}^i_{D\delta_e}, \bar{a}^i_{D0}\right)$ and $\underline{\omega}_{Di} = \left(\underline{a}^i_{D\alpha}, \underline{a}^i_{Dq}, \underline{a}^i_{D\delta_e}, \underline{a}^i_{D0}\right)$ as follows:

$$v^i_{D\mu} = \frac{\bar{a}^i_{D\mu} + \underline{a}^i_{D\mu}}{2}, \quad v^i_{Dq} = \frac{\bar{a}^i_{Dq} + \underline{a}^i_{Dq}}{2}, \quad v^i_{D\delta_e} = \frac{\bar{a}^i_{D\delta_e} + \underline{a}^i_{D\delta_e}}{2}, \quad v^i_{D0} = \frac{\bar{a}^i_{D0} + \underline{a}^i_{D0}}{2},$$

and

$$s^i_{D\mu} = \frac{\bar{a}^i_{D\mu} - \underline{a}^i_{D\mu}}{2}, \quad s^i_{Dq} = \frac{\bar{a}^i_{Dq} - \underline{a}^i_{Dq}}{2}, \quad s^i_{D\delta_e} = \frac{\bar{a}^i_{D\delta_e} - \underline{a}^i_{D\delta_e}}{2}, \quad s^i_{D0} = \frac{\bar{a}^i_{D0} - \underline{a}^i_{D0}}{2}.$$

Hence, consequent of rule $R^i_D$ is expressed as $C^i_D = \left[C^i_{Dl}, C^i_{Dr}\right]$.

$$C^i_{D_l} = v^i_{D_\alpha}\alpha_k + v^i_{D_q}q_k + v^i_{D_{\delta_\varepsilon}}\delta_{ek} + v^i_{D_0} - s^i_{D_\alpha}|\alpha_k| - s^i_{D_q}|q_k| - s^i_{D_{\delta_\varepsilon}}|\delta_{ek}| - s^i_{D_0} \tag{7.15a}$$

$$C^i_{D_r} = v^i_{D_\alpha}\alpha_k + v^i_{D_q}q_k + v^i_{D_{\delta_\varepsilon}}\delta_{ek} + v^i_{D_0} + s^i_{D_\alpha}|\alpha_k| + s^i_{D_q}|q_k| + s^i_{D_{\delta_\varepsilon}}|\delta_{ek}| + s^i_{D_0} \tag{7.15b}$$

The calculations used for modeling the drag force coefficient are further applied to develop the GT2 T-S FLS model. In the design, a triangular Secondary Membership Function (SMF) is selected, which can be represented through $\alpha$-plane decomposition. The base of the triangular SMF for each input is defined as

$$\left(\bar{\mu}_{\tilde{D}}(\alpha) - \underline{\mu}_{\tilde{D}}(\alpha)\right), \left(\bar{\mu}_{\tilde{D}}(q) - \underline{\mu}_{\tilde{D}}(q)\right), \left(\bar{\mu}_{\tilde{D}}(\delta_e) - \underline{\mu}_{\tilde{D}}(\delta_e)\right)$$

and its apex location is obtained as [14].

$$\text{apex}(\alpha) = \underline{\mu}_{\tilde{D}}(\alpha) + w_D\left[\bar{\mu}_{\tilde{D}}(\alpha) - \underline{\mu}_{\tilde{D}}(\alpha)\right], \tag{7.16a}$$

$$\text{apex}(q) = \underline{\mu}_{\tilde{D}}(q) + w_D\left[\bar{\mu}_{\tilde{D}}(q) - \underline{\mu}_{\tilde{D}}(q)\right], \tag{7.16b}$$

$$\text{apex}(\delta_e) = \underline{\mu}_{\tilde{D}}(\delta_e) + w_D\left[\bar{\mu}_{\tilde{D}}(\delta_e) - \underline{\mu}_{\tilde{D}}(\delta_e)\right], \tag{7.16c}$$

where $w_D \in [0, 1]$ work as design parameter. The $\alpha$-cuts of triangle SMFs for the first input parameter are given as,

$$\begin{cases} \tilde{D}_\alpha(\alpha) = \left[a_{\tilde{D}\alpha}(\alpha), b_{\tilde{D}\alpha}(\alpha)\right] \\ a_{\tilde{D}\alpha}(\alpha) = \underline{\mu}_{\tilde{D}}(\alpha) + w_D\left[\overline{\mu}_{\tilde{D}}(\alpha) - \underline{\mu}_{\tilde{D}}(\alpha)\right]\alpha \\ b_{\tilde{D}\alpha}(\alpha) = \overline{\mu}_{\tilde{D}}(\alpha) + (1-w_D)\left[\overline{\mu}_{\tilde{D}}(\alpha) - \underline{\mu}_{\tilde{D}}(\alpha)\right]\alpha \end{cases} \tag{7.17a}$$

Similarly, $\alpha$ cuts of triangle SMFs for the second and third input parameters $q$ and $\delta_e$ are given as (7.17b) and (7.17c).

$$\begin{cases} \tilde{D}_\alpha(q) = \left[a_{\tilde{D}\alpha}(q), b_{\tilde{D}\alpha}(q)\right] \\ a_{\tilde{D}\alpha}(q) = \underline{\mu}_{\tilde{D}}(q) + w_D\left[\overline{\mu}_{\tilde{D}}(q) - \underline{\mu}_{\tilde{D}}(q)\right]\alpha \\ b_{\tilde{D}\alpha}(q) = \overline{\mu}_{\tilde{D}}(q) + (1-w_D)\left[\overline{\mu}_{\tilde{D}}(q) - \underline{\mu}_{\tilde{D}}(q)\right]\alpha \end{cases} \tag{7.17b}$$

$$\begin{cases} \tilde{D}_\alpha(\delta_e) = \left[a_{\tilde{D}\alpha}(\delta_e), b_{\tilde{D}\alpha}(\delta_e)\right] \\ a_{\tilde{D}\alpha}(\delta_e) = \underline{\mu}_{\tilde{D}}(\delta_e) + w_D\left[\overline{\mu}_{\tilde{D}}(\delta_e) - \underline{\mu}_{\tilde{D}}(\delta_e)\right]\alpha \\ b_{\tilde{D}\alpha}(\delta_e) = \overline{\mu}_{\tilde{D}}(\delta_e) + (1-w_D)\left[\overline{\mu}_{\tilde{D}}(\delta_e) - \underline{\mu}_{\tilde{D}}(\delta_e)\right]\alpha \end{cases} \tag{7.17c}$$

The $\alpha$-cut decomposition for each rule-antecedent of type-1 vertical slice (when $[\alpha\ q\ \delta_e] = [\alpha', q', \delta_e']$) $\tilde{D}^i_\alpha(\alpha')$, $\tilde{D}^i_\alpha(q')$ and $\tilde{D}^i_\alpha(\delta_e')$ is obtained as

$$\tilde{D}^i_\alpha(\alpha') = \sup_{\forall\alpha\in[0,1]} \alpha/\left[a^i_{\tilde{D}\alpha}(\alpha'), b^i_{\tilde{D}\alpha}(\alpha')\right], \tag{7.18a}$$

$$\tilde{D}^i_\alpha(q') = \sup_{\forall\alpha\in[0,1]} \alpha/\left[a^i_{\tilde{D}\alpha}(q'), b^i_{\tilde{D}\alpha}(q')\right], \tag{7.18b}$$

$$\tilde{D}^i_\alpha(\delta_e') = \sup_{\forall\alpha\in[0,1]} \alpha/\left[a^i_{\tilde{D}\alpha}(\delta_e'), b^i_{\tilde{D}\alpha}(\delta_e')\right]. \tag{7.18c}$$

The $\alpha$ level firing interval, $\tilde{D}^i_\alpha[\alpha'\ q'\ \delta_e']$ is as (7.19)

$$D_\alpha : \begin{cases} \tilde{D}^i_\alpha[\alpha'\ q'\ \delta_e'] = [\underline{d}^i_\alpha[\alpha'\ q'\ \delta_e'], \overline{d}^i_\alpha[\alpha'\ q'\ \delta_e']] \\ \underline{d}^i_\alpha[\alpha'\ q'\ \delta_e'] = \min(a^i_{\tilde{D}\alpha}(\alpha'), a^i_{\tilde{D}\alpha}(q'), a^i_{\tilde{D}\alpha}(\delta_e')) \\ \overline{d}^i_\alpha[\alpha'\ q'\ \delta_e'] = \min(b^i_{\tilde{D}\alpha}(\alpha'), b^i_{\tilde{D}\alpha}(q'), b^i_{\tilde{D}\alpha}(\delta_e')) \end{cases} \tag{7.19}$$

As noted in [32], the design of the IT3 FLS employs both primary and secondary Footprints of Uncertainty (FOU). The secondary FOU is generated by introducing interval uncertainty in the secondary membership function, represented as $\tilde{\alpha}\ [\overline{\alpha},\ \underline{\alpha}]$.

The $\overline{\alpha}$ and $\underline{\alpha}$ planes representation of GT2 FS as presented in [32] is further used for the design of the IT3 T-S Drag coefficient model. The cut points on lower SMF ($\underline{\alpha}$) and upper SMF ($\overline{\alpha}$) are represented as $\underline{\alpha}/[a_{\mathrm{D}\underline{\alpha}}(\alpha'), b_{\mathrm{D}\underline{\alpha}}(\alpha')]$, $\underline{\alpha}/[a_{\mathrm{D}\underline{\alpha}}(q'), b_{\mathrm{D}\underline{\alpha}}(q')]$, $\underline{\alpha}/[a_{\mathrm{D}\underline{\alpha}}(\delta_e'), b_{\mathrm{D}\underline{\alpha}}(\delta_e')]$ and $\overline{\alpha}/[a_{\tilde{\mathrm{D}}\overline{\alpha}}(\alpha'), b_{\tilde{\mathrm{D}}\overline{\alpha}}(\alpha')]$, $\overline{\alpha}/[a_{\tilde{\mathrm{D}}\overline{\alpha}}(q'), b_{\tilde{\mathrm{D}}\overline{\alpha}}(q')]$, $\overline{\alpha}/[a_{\tilde{\mathrm{D}}\overline{\alpha}}(\delta_e'), b_{\tilde{\mathrm{D}}\overline{\alpha}}(\delta_e')]$, respectively. The tertiary grade of membership is unity.

Each $\bar{\alpha}$ and $\underline{\alpha}$ are discretize for $\lambda = 1, ....\rho$ levels. Then by taking the union of all horizontal slices $\tilde{\alpha_1}, \tilde{\alpha_2}, ....\tilde{\alpha_\lambda}$ the IT3 MF are generated, where $\tilde{\alpha_\lambda} \subseteq [\underline{\alpha_\lambda}\ \bar{\alpha_\lambda}]$ . The horizontal slice of IT3 MF at level $\tilde{\alpha_\lambda}$ signifies an inter-val type-2 MF where secondary grade is between $[\underline{\alpha_\lambda}\ \bar{\alpha_\lambda}]$ .

The technique for IT3 T-S FLS design is as follows:

1. Select the $\alpha$- cuts number $\rho$ for all the vertical slices of $\tilde{D}$ and discretize the values of $\alpha_\lambda$ as $\alpha_1, \alpha_2, .....\alpha_\rho$ for $\lambda = 1, ..., \rho$ levels. Signify the uncertainty in secondary grades as $[\underline{\alpha}_\lambda\ \bar{\alpha}_\lambda]$ where $\underline{\alpha}_\lambda = \alpha_\lambda^\Delta$ and $\overline{\alpha}_\lambda = \alpha_\lambda^{1/\Delta}$ i.e., $\underline{\alpha}_\lambda$ as $\underline{\alpha}_1, \underline{\alpha}_2, \ldots, \underline{\alpha}_\rho$ and $\overline{\alpha}_\lambda$ as $\overline{\alpha}_1, \overline{\alpha}_2, \ldots, \overline{\alpha}_\rho$.
   The parameter $\Delta \geq 1$ is recognized as fuzziness parameter and it shows the vertical slice uncertainty at $[\alpha_i\ \ q_i\ \ \delta_{ei}] = [\alpha_i'\ \ q_i'\ \ \delta_{ei}']$.
2. For each $[\alpha_k\ q_k\ \delta_{ek}]$, over the support of $\tilde{D}$, compute $[a^i_{\tilde{D}\underline{\alpha}_\lambda}(\alpha_k), b^i_{\tilde{D}\underline{\alpha}_\lambda}(\alpha_k)]$, $[a^i_{\tilde{D}\underline{\alpha}_\lambda}(q_k), b^i_{\tilde{D}\underline{\alpha}_\lambda}(q_k)]$, $[a^i_{\tilde{D}\underline{\alpha}_\lambda}(\delta_{ek}), b^i_{\tilde{D}\underline{\alpha}_\lambda}(\delta_{ek})]$ and $[a^i_{\tilde{D}\overline{\alpha}_\lambda}(\alpha_k), b^i_{\tilde{D}\overline{\alpha}_\lambda}(\alpha_k)]$, $[a^i_{\tilde{D}\overline{\alpha}_\lambda}(q_k), b^i_{\tilde{D}\overline{\alpha}_\lambda}(q_k)]$, $[a^i_{\tilde{D}\overline{\alpha}_\lambda}(\delta_{ek}), b^i_{\tilde{D}\overline{\alpha}_\lambda}(\delta_{ek})]$ for $\lambda = 1, \ldots, \rho$ as.

$$\begin{cases} a^i_{\tilde{D}\underline{\alpha}_\lambda}(\alpha_k) = \underline{\mu}_{\tilde{D}_i}(\alpha_k) + w_D\Big[\overline{\mu}_{\tilde{D}_i}(\alpha_k) - \underline{\mu}_{\tilde{D}_i}(\alpha_k)\Big]\underline{\alpha}_\lambda, \\ b_{\tilde{D}_{\underline{\alpha}_\lambda}}(\alpha_k) = \overline{\mu}_{\tilde{D}_i}(\alpha_k) + (1 - w_D)\Big[\overline{\mu}_{\tilde{D}_i}(\alpha_k) - \underline{\mu}_{\tilde{D}_i}(\alpha_k)\Big]\underline{\alpha}_\lambda \end{cases} \tag{7.20a}$$

$$\begin{cases} a^i_{\tilde{D}\underline{\alpha}_\lambda}(q_k) = \underline{\mu}_{\tilde{D}_i}(q_k) + w_D\Big[\overline{\mu}_{\tilde{D}_i}(q_k) - \underline{\mu}_{\tilde{D}_i}(q_k)\Big]\underline{\alpha}_\lambda, \\ b_{\tilde{D}_{\underline{\alpha}_\lambda}}(q_k) = \overline{\mu}_{\tilde{D}_i}(q_k) + (1 - w_D)\Big[\overline{\mu}_{\tilde{D}_i}(q_k) - \underline{\mu}_{\tilde{D}_i}(q_k)\Big]\underline{\alpha}_\lambda \end{cases} \tag{7.20b}$$

$$\begin{cases} a^i_{\tilde{D}\underline{\alpha}_\lambda}(\delta_{ek}) = \underline{\mu}_{\tilde{D}_i}(\delta_{ek}) + w_D\Big[\overline{\mu}_{\tilde{D}_i}(\delta_{ek}) - \underline{\mu}_{\tilde{D}_i}(\delta_{ek})\Big]\underline{\alpha}_\lambda, \\ b_{\tilde{D}_{\underline{\alpha}_\lambda}}(\delta_{ek}) = \overline{\mu}_{\tilde{D}_i}(\delta_{ek}) + (1 - w_D)\Big[\overline{\mu}_{\tilde{D}_i}(\delta_{ek}) - \underline{\mu}_{\tilde{D}_i}(\delta_{ek})\Big]\underline{\alpha}_\lambda \end{cases} \tag{7.20c}$$

$$\begin{cases} a^i_{\tilde{D}\bar{\alpha}_\lambda}(\alpha_k) = \underline{\mu}_{\tilde{D}_i}(\alpha_k) + w_D\Big[\overline{\mu}_{\tilde{D}_i}(\alpha_k) - \underline{\mu}_{\tilde{D}_i}(\alpha_k)\Big]\bar{\alpha}_\lambda, \\ b_{\tilde{D}_{\bar{\alpha}_\lambda}}(\alpha_k) = \overline{\mu}_{\tilde{D}_i}(\alpha_k) + (1 - w_D)\Big[\overline{\mu}_{\tilde{D}_i}(\alpha_k) - \underline{\mu}_{\tilde{D}_i}(\alpha_k)\Big]\bar{\alpha}_\lambda \end{cases} \tag{7.21a}$$

$$\begin{cases} a^i_{\tilde{D}\bar{\alpha}_\lambda}(q_k) = \underline{\mu}_{\tilde{D}_i}(q_k) + w_D\Big[\overline{\mu}_{\tilde{D}_i}(q_k) - \underline{\mu}_{\tilde{D}_i}(q_k)\Big]\bar{\alpha}_\lambda, \\ b_{\tilde{D}_{\bar{\alpha}_\lambda}}(q_k) = \overline{\mu}_{\tilde{D}_i}(q_k) + (1 - w_D)\Big[\overline{\mu}_{\tilde{D}_i}(q_k) - \underline{\mu}_{\tilde{D}_i}(q_k)\Big]\bar{\alpha}_\lambda \end{cases} \tag{7.21b}$$

$$\begin{cases} a^i_{\tilde{D}\bar{\alpha}_\lambda}(\delta_{ek}) = \underline{\mu}_{\tilde{D}_i}(\delta_{ek}) + w_D\Big[\overline{\mu}_{\tilde{D}_i}(\delta_{ek}) - \underline{\mu}_{\tilde{D}_i}(\delta_{ek})\Big]\bar{\alpha}_\lambda, \\ b_{\tilde{D}_{\bar{\alpha}_\lambda}}(\delta_{ek}) = \overline{\mu}_{\tilde{D}_i}(\delta_{ek}) + (1 - w_D)\Big[\overline{\mu}_{\tilde{D}_i}(\delta_{ek}) - \underline{\mu}_{\tilde{D}_i}(\delta_{ek})\Big]\bar{\alpha}_\lambda \end{cases} \tag{7.21c}$$

3. Determine the firing interval $\tilde{D}^i_{\underline{\alpha}_\lambda}[\alpha_k \quad q_k \quad \delta_{e_k}]$ and $\tilde{D}^i_{\overline{\alpha}_\lambda}[\alpha_k \quad q_k \quad \delta_{e_k}]$ for each level $\underline{\alpha}_\lambda$ and $\overline{\alpha}_\lambda$ corresponding to $i$th rule as

$$\begin{cases} \widetilde{D}^i_{\underline{\alpha}_\lambda}[\alpha_k, q_k, \delta_{ek}] \equiv \left[\underline{d}^i_{\underline{\alpha}_\lambda}[\alpha_k, q_k, \delta_{ek}], \overline{d}^i_{\underline{\alpha}_\lambda}[\alpha_k, q_k, \delta_{ek}]\right], \\ \underline{d}^i_{\underline{\alpha}_\lambda}[\alpha_k, q_k, \delta_{ek}] = \min\left(a^i_{\widetilde{D}\underline{\alpha}_\lambda}(\alpha_k), a^i_{\widetilde{D}\underline{\alpha}_\lambda}(q_k), a^i_{\widetilde{D}\underline{\alpha}_\lambda}(\delta_{ek})\right), \\ \overline{d}^i_{\underline{\alpha}_\lambda}[\alpha_k, q_k, \delta_{ek}] = \min\left(b^i_{\widetilde{D}\underline{\alpha}_\lambda}(\alpha_k), b^i_{\widetilde{D}\underline{\alpha}_\lambda}(q_k), b^i_{\widetilde{D}\underline{\alpha}_\lambda}(\delta_{ek})\right) \end{cases} \tag{7.22a}$$

$$\begin{cases} \widetilde{D}^i_{\overline{\alpha}_\lambda}[\alpha_k, q_k, \delta_{ek}] \equiv \left[\underline{d}^i_{\overline{\alpha}_\lambda}[\alpha_k, q_k, \delta_{ek}], \overline{d}^i_{\overline{\alpha}_\lambda}[\alpha_k, q_k, \delta_{ek}]\right], \\ \underline{d}^i_{\overline{\alpha}_\lambda}[\alpha_k, q_k, \delta_{ek}] = \min\left(a^i_{\widetilde{D}\overline{\alpha}_\lambda}(\alpha_k), a^i_{\widetilde{D}\overline{\alpha}_\lambda}(q_k), a^i_{\widetilde{D}\overline{\alpha}_\lambda}(\delta_{ek})\right), \\ \overline{d}^i_{\overline{\alpha}_\lambda}[\alpha_k, q_k, \delta_{ek}] = \min\left(b^i_{\widetilde{D}\overline{\alpha}_\lambda}(\alpha_k), b^i_{\widetilde{D}\overline{\alpha}_\lambda}(q_k), b^i_{\widetilde{D}\overline{\alpha}_\lambda}(\delta_{ek})\right) \end{cases} \tag{7.22b}$$

4. Estimate the $\underline{\alpha}_\lambda$ and $\overline{\alpha}_\lambda$ level $m$-$n$ outputs $C_{Dmn,\underline{\alpha}_\lambda}([\alpha_k \quad q_k \quad \delta_{ek}])$ and $C_{Dmn,\overline{\alpha}_\lambda}([\alpha_k\ q_k\ \delta_{ek}])$ at each $\underline{\alpha}_1, \underline{\alpha}_2, \ldots, \underline{\alpha}_\rho$ and $\overline{\alpha}_1, \overline{\alpha}_2, \ldots, \overline{\alpha}_\rho$ as [14, 41]

$$C_{Dmn,\underline{\alpha}_\lambda}([\alpha_k\ q_k\ \delta_{ek}]) = m\frac{\sum_{i=1}^{c} C^i_{Dl}\underline{d}^i_{\underline{\alpha}_\lambda}[\alpha_k\ q_k\ \delta_{ek}]}{\sum_{i=1}^{c}\underline{d}^i_{\underline{\alpha}_\lambda}[\alpha_k\ q_k\ \delta_{ek}]} + n\frac{\sum_{i=1}^{c} C^i_{Dr}\overline{d}^i_{\underline{\alpha}_\lambda}[\alpha_k\ q_k\ \delta_{ek}]}{\sum_{i=1}^{c}\overline{d}^i_{\underline{\alpha}_\lambda}[\alpha_k\ q_k\ \delta_{ek}]} \tag{7.23a}$$

$$C_{Dmn,\bar{\alpha}_\lambda}([\alpha_k\ q_k\ \delta_{ek}]) = m\frac{\sum_{i=1}^{c} C^i_{Dl}\underline{d}^i_{\bar{\alpha}_\lambda}[\alpha_k\ q_k\ \delta_{ek}]}{\sum_{i=1}^{c}\underline{d}^i_{\bar{\alpha}_\lambda}[\alpha_k\ q_k\ \delta_{ek}]} + n\frac{\sum_{i=1}^{c} C^i_{Dr}\overline{d}^i_{\bar{\alpha}_\lambda}[\alpha_k\ q_k\ \delta_{ek}]}{\sum_{i=1}^{c}\overline{d}^i_{\bar{\alpha}_\lambda}[\alpha_k\ q_k\ \delta_{ek}]} \tag{7.23b}$$

where $[C^i_{Dl}, C^i_{Dr}]$ denote the rule consequent and same is achieved from (7.15a)–(7.15b).

5. Defuzzify the $C_{Dmn,\underline{\alpha}_\lambda}([\alpha_k\ q_k\ \delta_{ek}])$ and $C_{Dmn,\overline{\alpha}_\lambda}([\alpha_k\ q_k\ \delta_{ek}])$ for $\underline{\alpha}_1, \underline{\alpha}_2, \ldots, \underline{\alpha}_\rho$ and $\overline{\alpha}_1, \overline{\alpha}_2, \ldots, \overline{\alpha}_\rho$, and create the Drag coefficient IT3 FLS as

$$C_{Dmn\text{-IT3}}([\alpha_k\ q_k\ \delta_{ek}]) = \frac{\sum_{\lambda=1}^{\rho}(\underline{\alpha}_\lambda C_{Dmn,\underline{\alpha}_\lambda}([\alpha_k\ q_k\ \delta_{ek}]) + \overline{\alpha}_\lambda C_{Dmn,\overline{\alpha}_\lambda}([\alpha_k\ q_k\ \delta_{ek}]))}{\sum_{\lambda=1}^{\rho}(\underline{\alpha}_\lambda + \overline{\alpha}_\lambda)} \tag{7.24}$$

6. The $m, n$ tuning parameters are optimized for Mean Square Error (MSE) minimization and are found by utilizing the algorithms offered in [41].

$$\text{MSE}_D = \frac{1}{N}\sum_{k=1}^{N}\left(C_{\text{Dmn-IT3}}([\alpha_k \quad q_k \quad \delta_{ek}]) - C_{Dk}\right)^2 \tag{7.25}$$

### 7.3.2 IT3 T-S Fuzzy Model for Lift Force Coefficient

The IT3 T-S fuzzy model for the lift coefficient is constructed using the data pairs

$$([\alpha_k, q_k, \delta e_k], C_{Lk}), \quad k = 1, 2, \ldots, N.$$

The input parameters are the same as those used in the drag coefficient model, while the output is the lift coefficient $C_L$. The formulation presented in Sect. 7.3.1 (drag coefficient model) can be directly adapted for the lift coefficient by modifying the notation. Similar to the drag model, the lift coefficient is represented using a hyperplane regression function as

$$\hat{C}_{L_k} = f^i\left([\alpha_k \; q_k \; \delta_{e_k}], \boldsymbol{\omega}_{Li}\right) = a^i_{L\alpha}\alpha_k + a^i_{Lq}q_k + a^i_{L\delta_e}\delta_{e_k} + a^i_{L0}, \tag{7.26}$$

where, $\boldsymbol{\omega}_{L_i} = \left[a^i_{L\alpha}, \; a^i_{Lq}, \; a^i_{L\delta_e}, \; a^i_{L0}\right]$ is the hyperplane model coefficient vector.

The upper and lower hyperplanes are obtained as

$$\overline{\boldsymbol{\omega}}_{L_i} = \left[\mathbf{X}^T\overline{\mathbf{P}}_{L_i}[\mathbf{X}]\right]^{-1}\mathbf{X}^T\overline{\mathbf{P}}_{L_i}\mathbf{C}_L, \tag{7.27}$$

$$\underline{\boldsymbol{\omega}}_{L_i} = \left[\mathbf{X}^T\underline{\mathbf{P}}_{L_i}\mathbf{X}\right]^{-1}\mathbf{X}^T\underline{\mathbf{P}}_{L_i}\mathbf{C}_L, \tag{7.28}$$

where partition matrices $\overline{\mathbf{P}}_{L_i}$ and $\underline{\mathbf{P}}_{L_i}$ are

$$\overline{\mathbf{P}}_{L_i} = \begin{bmatrix} \overline{u}_{Li1} & 0 & \cdots & 0 \\ 0 & \overline{u}_{Li2} & \cdots & 0 \\ \vdots & \vdots & \ddots & \vdots \\ 0 & 0 & \cdots & \overline{u}_{LiN} \end{bmatrix}, \quad \underline{\mathbf{P}}_{L_i} = \begin{bmatrix} \underline{u}_{Li1} & 0 & \cdots & 0 \\ 0 & \underline{u}_{Li2} & \cdots & 0 \\ \vdots & \vdots & \ddots & \vdots \\ 0 & 0 & \cdots & \underline{u}_{LiN} \end{bmatrix} \in R^{N\times N}.$$

Like the drag coefficient model, the hyperplane-shaped Gaussian MF is utilized for each input parameter. The upper and lower membership degrees for each input parameter are as follows:

For first input parameter $\alpha_k$:

$$\overline{\mu}_{\tilde{L}_i}(\alpha_k) = \exp\left(-\eta_L \cdot \frac{d_{Li\alpha_k}(\overline{\boldsymbol{\omega}}_{L_i})}{\max\left(\{d_{Li\alpha_k}(\overline{\boldsymbol{\omega}}_{L_i})\}, \; i = 1, \ldots, c\right)}\right), \tag{7.29a}$$

$$\underline{\mu}_{\tilde{L}_i}(\alpha_k) = \exp\left(-\eta_L \cdot \frac{d_{Li_{\alpha_k}}(\underline{\boldsymbol{\omega}}_{Li})}{\max(\{d_{Li_{\alpha_k}}(\underline{\boldsymbol{\omega}}_{Li}), i = 1, \ldots, c\})}\right), \tag{7.29b}$$

where, $d_{Li\alpha_k}(\bar{\boldsymbol{\omega}}_{Li})$ represent the distance of $\alpha_k$ from $i$th upper and $d_{Li\alpha_k}(\underline{\boldsymbol{\omega}}_{Li})$ is the distance of $\alpha_k$ from $i$th lower hyper planes. The parameter $\eta_L$ is a constant, which is used for tuning of upper and lower membership grades $\bar{\mu}_{\tilde{L}_i}(\alpha_k)$ and $\underline{\mu}_{\tilde{L}_i}(\alpha_k)$. These parameters regulate the fuzziness of the membership grades and are referred to as the antecedent membership function parameters. The parameters $d_{Li\alpha_k}(\bar{\boldsymbol{\omega}}_{Li})$ and $d_{Li\alpha_k}(\underline{\boldsymbol{\omega}}_{Li})$ are obtained as follows:

$$d_{Li\alpha_k}(\bar{\boldsymbol{\omega}}_{Li}) = \frac{|\alpha_k \cdot \bar{\boldsymbol{\omega}}_{Li}|}{\|\bar{\boldsymbol{\omega}}_{Li}\|} \quad \text{and} \quad d_{Li\alpha_k}(\underline{\boldsymbol{\omega}}_{Li}) = \frac{|\alpha_k \cdot \underline{\boldsymbol{\omega}}_{Li}|}{\|\underline{\boldsymbol{\omega}}_{Li}\|}$$

For second input parameter $q_k$:

$$\bar{\mu}_{\tilde{L}_i}(q_k) = \exp\left(-\eta_L \cdot \frac{d_{Liq_k}(\bar{\boldsymbol{\omega}}_{Li})}{\max(\{d_{Liq_k}(\bar{\boldsymbol{\omega}}_{Li}), i = 1, \ldots, c\})}\right), \tag{7.30a}$$

$$\underline{\mu}_{\tilde{L}_i}(q_k) = \exp\left(-\eta_L \cdot \frac{d_{Liq_k}(\underline{\boldsymbol{\omega}}_{Li})}{\max(\{d_{Liq_k}(\underline{\boldsymbol{\omega}}_{Li}), i = 1, \ldots, c\})}\right), \tag{7.30b}$$

where, $d_{Liq_k}(\bar{\omega}_{Li})$ is the distance of $q_k$ from $i$th upper and $d_{Liq_k}(\underline{\omega}_{Li})$ is the distance of $q_k$ from $i$th lower hyper planes

$$d_{Liq_k}(\bar{\boldsymbol{\omega}}_{Li}) = \frac{|q_k \cdot \bar{\boldsymbol{\omega}}_{Li}|}{\|\bar{\boldsymbol{\omega}}_{Li}\|} \quad \text{and} \quad d_{Liq_k}(\underline{\boldsymbol{\omega}}_{Li}) = \frac{|q_k \cdot \underline{\boldsymbol{\omega}}_{Li}|}{\|\underline{\boldsymbol{\omega}}_{Li}\|}$$

For third input parameter $\delta_{ek}$:

$$\bar{\mu}_{\tilde{L}_i}(\delta_{ek}) = \exp\left(-\eta_L \cdot \frac{d_{Li\delta_{ek}}(\bar{\boldsymbol{\omega}}_{Li})}{\max(\{d_{Li\delta_{ek}}(\bar{\boldsymbol{\omega}}_{Li}), i = 1, \ldots, c\})}\right), \tag{7.31a}$$

$$\underline{\mu}_{\tilde{L}_i}(\delta_{ek}) = \exp\left(-\eta_L \cdot \frac{d_{Li\delta_{ek}}(\underline{\boldsymbol{\omega}}_{Li})}{\max(\{d_{Li\delta_{ek}}(\underline{\boldsymbol{\omega}}_{Li}), i = 1, \ldots, c\})}\right), \tag{7.31b}$$

where, $d_{Li\delta_{ek}}(\bar{\omega}_{Li})$ is the distance of $\delta_{ek}$ from $i$th upper and $d_{Li\delta_{ek}}(\underline{\omega}_{Li})$ is the distance of $\delta_{ek}$ from $i$th lower hyper planes

$$d_{Li_{\delta_{ek}}}(\bar{\boldsymbol{\omega}}_{Li}) = \frac{|\delta_{ek} \cdot \bar{\boldsymbol{\omega}}_{Li}|}{\|\bar{\boldsymbol{\omega}}_{Li}\|} \quad \text{and} \quad d_{Li_{\delta_{ek}}}(\underline{\boldsymbol{\omega}}_{Li}) = \frac{|\delta_{ek} \cdot \underline{\boldsymbol{\omega}}_{Li}|}{\|\underline{\boldsymbol{\omega}}_{Li}\|}$$

An IT2 T-S FLS for modeling of the Lift parameter is defined by the following IF-THEN fuzzy rules ($i$th rule)

$$R_L^i : \text{ IF } \alpha \text{ is } \tilde{L}_\alpha^i \text{ and } q \text{ is } \tilde{L}_q^i \text{ and } \delta_e \text{ is } \tilde{L}_{\delta_e}^i$$

THEN

$$C_L^i(\alpha, q, \delta_e) = C_{L\alpha}^i\, \alpha + C_{Lq}^i\, q + C_{L\delta_e}^i\, \delta_e + C_{L0}^i$$

Where $\tilde{L}_\alpha^i, \tilde{L}_q^i, \tilde{L}_{\delta_e}^i$ are fuzzy model terms for the lift coefficient model related to each input parameter.

The interval type consequent parameters $\left(C_{L\alpha}^i, C_{Lq}^i, C_{L\delta_e}^i, C_{L0}^i\right)$ of IT2 T-S FLS are assessed from upper and lower hyper-planes coefficients $\bar{\omega}_{Li}$ and $\underline{\omega}_{Li}$ . Like the drag coefficient model, the consequence of rule $R_L^i$ is expressed as $C_L^i = [C_{Ll}^i, C_{Lr}^i]$ and also found from lower and upper hyper-planes coefficient vectors.

The base of SMF (triangular type) for each input equals $\left(\overline{\mu}_{\tilde{L}}(\alpha) - \underline{\mu}_{\tilde{L}}(\alpha)\right)$, $\left(\overline{\mu}_{\tilde{L}}(q) - \underline{\mu}_{\tilde{L}}(q)\right)$, $\left(\overline{\mu}_{\tilde{L}}(\delta_e) - \underline{\mu}_{\tilde{L}}(\delta_e)\right)$ and its apex location is obtained as [14].

$$\text{apex}(\alpha) = \underline{\mu}_{\tilde{L}}(\alpha) + w_L \left[\overline{\mu}_{\tilde{L}}(\alpha) - \underline{\mu}_{\tilde{L}}(\alpha)\right], \tag{7.32a}$$

$$\text{apex}(q) = \underline{\mu}_{\tilde{L}}(q) + w_L \left[\overline{\mu}_{\tilde{L}}(q) - \underline{\mu}_{\tilde{L}}(q)\right], \tag{7.32b}$$

$$\text{apex}(\delta_e) = \underline{\mu}_{\tilde{L}}(\delta_e) + w_L \left[\overline{\mu}_{\tilde{L}}(\delta_e) - \underline{\mu}_{\tilde{L}}(\delta_e)\right], \tag{7.32c}$$

where $w_L \in [0, 1]$ signifies the design parameter. The $\alpha$-cuts of triangle SMFs for the first input parameter are given as:

$$\begin{cases} \tilde{L}_\alpha(\alpha) = \left[a_{\tilde{L}\alpha}(\alpha), b_{\tilde{L}\alpha}(\alpha)\right] \\ a_{\tilde{L}\alpha}(\alpha) = \underline{\mu}_{\tilde{L}}(\alpha) + w_L \left[\overline{\mu}_{\tilde{L}}(\alpha) - \underline{\mu}_{\tilde{L}}(\alpha)\right] \alpha \\ b_{\tilde{L}\alpha}(\alpha) = \overline{\mu}_{\tilde{L}}(\alpha) + (1 - w_L) \left[\overline{\mu}_{\tilde{L}}(\alpha) - \underline{\mu}_{\tilde{L}}(\alpha)\right] \alpha \end{cases} \tag{7.33a}$$

Similarly, $\alpha$ cuts of triangle SMFs for the second and third input parameters $q$ and $\delta_e$ are given as (7.33b) and (7.33c).

$$\begin{cases} \tilde{L}_\alpha(q) = \left[a_{\tilde{L}\alpha}(q), b_{\tilde{L}\alpha}(q)\right] \\ a_{\tilde{L}\alpha}(q) = \underline{\mu}_{\tilde{L}}(q) + w_L \left[\overline{\mu}_{\tilde{L}}(q) - \underline{\mu}_{\tilde{L}}(q)\right] \alpha \\ b_{\tilde{L}\alpha}(q) = \overline{\mu}_{\tilde{L}}(q) + (1 - w_L) \left[\overline{\mu}_{\tilde{L}}(q) - \underline{\mu}_{\tilde{L}}(q)\right] \alpha \end{cases} \tag{7.33b}$$

$$\begin{cases} \tilde{L}_\alpha(\delta_e) = \left[a_{\tilde{L}\alpha}(\delta_e), b_{\tilde{L}\alpha}(\delta_e)\right] \\ a_{\tilde{L}\alpha}(\delta_e) = \underline{\mu}_{\tilde{L}}(\delta_e) + w_L \left[\overline{\mu}_{\tilde{L}}(\delta_e) - \underline{\mu}_{\tilde{L}}(\delta_e)\right] \alpha \\ b_{\tilde{L}\alpha}(\delta_e) = \overline{\mu}_{\tilde{L}}(\delta_e) + (1 - w_L) \left[\overline{\mu}_{\tilde{L}}(\delta_e) - \underline{\mu}_{\tilde{L}}(\delta_e)\right] \alpha \end{cases} \tag{7.33c}$$

The $\alpha$-cut decomposition for each rule-antecedent of type-1 vertical slice (when $[\alpha\ q\ \delta_e] = [\alpha', q', \delta_e']$) $\tilde{L}^i_\alpha(\alpha')$, $\tilde{L}^i_\alpha(q')$ and $\tilde{L}^i_\alpha(\delta_e')$ is obtained as

$$\tilde{L}^i_\alpha(\alpha') = \sup_{\forall \alpha \in [0,1]} \alpha / \left[a^i_{\tilde{L}\alpha}(\alpha'), b^i_{\tilde{L}\alpha}(\alpha')\right] \tag{7.34a}$$

$$\tilde{L}^i_\alpha(q') = \sup_{\forall \alpha \in [0,1]} \alpha / \left[a^i_{\tilde{L}\alpha}(q'), b^i_{\tilde{L}\alpha}(q')\right] \tag{7.34b}$$

$$\tilde{L}^i_\alpha(\delta_e') = \sup_{\forall \alpha \in [0,1]} \alpha / \left[a^i_{\tilde{L}\alpha}(\delta_e'), b^i_{\tilde{L}\alpha}(\delta_e')\right] \tag{7.34c}$$

The $\alpha$ level firing interval, $\tilde{L}^i_\alpha[\alpha'\ q'\ \delta_e']$ is as (7.35)

$$\begin{aligned} \tilde{L}^i_\alpha[\alpha'\ q'\ \delta_e'] &= [\underline{l}^i_\alpha[\alpha'\ q'\ \delta_e'], \overline{l}^i_\alpha[\alpha'\ q'\ \delta_e']] \\ \underline{l}^i_\alpha[\alpha'\ q'\ \delta_e'] &= \min(a^i_{\tilde{L}\alpha}(\alpha'), a^i_{\tilde{L}\alpha}(q'), a^i_{\tilde{L}\alpha}(\delta_e')) \\ \overline{l}^i_\alpha[\alpha'\ q'\ \delta_e'] &= \min(b^i_{\tilde{L}\alpha}(\alpha'), b^i_{\tilde{L}\alpha}(q'), b^i_{\tilde{L}\alpha}(\delta_e')) \end{aligned} \tag{7.35}$$

The lower SMF ($\underline{\alpha}$) and upper SMF ($\overline{\alpha}$) cut points are represented as $\underline{\alpha}/[a_{\tilde{L}\underline{\alpha}}(\alpha'), b_{\tilde{L}\underline{\alpha}}(\alpha')]$, $\underline{\alpha}/[a_{\tilde{L}\underline{\alpha}}(q'), b_{\tilde{L}\underline{\alpha}}(q')]$, $\underline{\alpha}/[a_{\tilde{L}\underline{\alpha}}(\delta_e'), b_{\tilde{L}\underline{\alpha}}(\delta_e')]$ and $\overline{\alpha}/[a_{\tilde{L}\overline{\alpha}}(\alpha'), b_{\tilde{L}\overline{\alpha}}(\alpha')]$, $\overline{\alpha}/[a_{\tilde{L}\overline{\alpha}}(q'), b_{\tilde{L}\overline{\alpha}}(q')]$, $\overline{\alpha}/[a_{\tilde{L}\overline{\alpha}}(\delta_e'), b_{\tilde{L}\overline{\alpha}}(\delta_e')]$, respectively. Like the drag coefficient model, these cut points are estimated for discretized $\underline{\alpha}_\lambda$ and $\overline{\alpha}_\lambda$ levels similar to (7.20a)–(7.20c) and (7.21a)–(7.21c) for $\lambda = 1, \ldots, \rho$.

Further the firing interval $\underline{\tilde{L}}^i_{\underline{\alpha}_\lambda}[\alpha_k\quad q_k\quad \delta_{e_k}]$ and $\tilde{L}_{\overline{\alpha}_\lambda}[\alpha_k\quad q_k\quad \delta_{e_k}]$ for each $\underline{\alpha}_\lambda$ and $\overline{\alpha}_\lambda$ level corresponding to $i$th rule are as

$$\begin{cases} \tilde{L}^i_{\underline{\alpha}_\lambda}[\alpha_k, q_k, \delta_{ek}] \equiv \left[\underline{l}^i_{\underline{\alpha}_\lambda}[\alpha_k, q_k, \delta_{ek}], \overline{l}^i_{\underline{\alpha}_\lambda}[\alpha_k, q_k, \delta_{ek}]\right], \\ \underline{l}^i_{\underline{\alpha}_\lambda}[\alpha_k, q_k, \delta_{ek}] = \min\left(a^i_{\tilde{L}\underline{\alpha}_\lambda}(\alpha_k), a^i_{\tilde{L}\underline{\alpha}_\lambda}(q_k), a^i_{\tilde{L}\underline{\alpha}_\lambda}(\delta_{ek})\right), \\ \overline{l}^i_{\underline{\alpha}_\lambda}[\alpha_k, q_k, \delta_{ek}] = \min\left(b^i_{\tilde{L}_{\underline{\alpha}_\lambda}}(\alpha_k), b^i_{\tilde{L}\underline{\alpha}_\lambda}(q_k), b^i_{\tilde{L}\underline{\alpha}_\lambda}(\delta_{ek})\right) \end{cases} \tag{7.36a}$$

$$\begin{cases} \tilde{L}^i_{\overline{\alpha}_\lambda}[\alpha_k, q_k, \delta_{ek}] \equiv \left[\underline{l}^i_{\overline{\alpha}_\lambda}[\alpha_k, q_k, \delta_{ek}], \overline{l}^i_{\overline{\alpha}_\lambda}[\alpha_k, q_k, \delta_{ek}]\right], \\ \underline{l}^i_{\overline{\alpha}_\lambda}[\alpha_k, q_k, \delta_{ek}] = \min\left(a^i_{\tilde{L}\overline{\alpha}_\lambda}(\alpha_k), a^i_{\tilde{L}\overline{\alpha}_\lambda}(q_k), a^i_{\tilde{L}\overline{\alpha}_\lambda}(\delta_{ek})\right), \\ \overline{l}^i_{\overline{\alpha}_\lambda}[\alpha_k, q_k, \delta_{ek}] = \min\left(b^i_{\tilde{L}\overline{\alpha}_\lambda}(\alpha_k), b^i_{\tilde{L}\overline{\alpha}_\lambda}(q_k), b^i_{\tilde{L}\overline{\alpha}_\lambda}(\delta_{ek})\right) \end{cases} \tag{7.36b}$$

In the next step $\underline{\alpha}_\lambda$ and $\overline{\alpha}_\lambda$ level $m$-$n$ outputs $C_{Lmn,\underline{\alpha}_\lambda}([\alpha_k\ q_k\ \delta_{ek}])$ and $C_{Lmn,\overline{\alpha}_\lambda}([\alpha_k\ q_k\ \delta_{ek}])$ at each $\underline{\alpha}_1, \underline{\alpha}_2, \ldots, \underline{\alpha}_\rho$ and $\overline{\alpha}_1, \overline{\alpha}_2, \ldots, \overline{\alpha}_\rho$ as,

$$C_{Lmn,\underline{\alpha}_\lambda}([\alpha_k\ q_k\ \delta_{ek}]) = m\frac{\sum_{i=1}^{c} C_{Ll}^{i}\underline{l}_{\underline{\alpha}_\lambda}^{i}[\alpha_k\ q_k\ \delta_{ek}]}{\sum_{i=1}^{c} \underline{l}_{\underline{\alpha}_\lambda}^{i}[\alpha_k\ q_k\ \delta_{ek}]} + n\frac{\sum_{i=1}^{c} C_{Lr}^{i}\overline{l}_{\underline{\alpha}_\lambda}^{i}[\alpha_k\ q_k\ \delta_{ek}]}{\sum_{i=1}^{c} \overline{l}_{\underline{\alpha}_\lambda}^{i}[\alpha_k\ q_k\ \delta_{ek}]} \tag{7.37a}$$

$$C_{Lmn,\bar{\alpha}_\lambda}([\alpha_k\ q_k\ \delta_{ek}]) = m\frac{\sum_{i=1}^{c} C_{Ll}^{i}\underline{l}_{\bar{\alpha}_\lambda}^{i}[\alpha_k\ q_k\ \delta_{ek}]}{\sum_{i=1}^{c} \underline{l}_{\bar{\alpha}_\lambda}^{i}[\alpha_k\ q_k\ \delta_{ek}]} + n\frac{\sum_{i=1}^{c} C_{Lr}^{i}\overline{l}_{\bar{\alpha}_\lambda}^{i}[\alpha_k\ q_k\ \delta_{ek}]}{\sum_{i=1}^{c} \overline{l}_{\bar{\alpha}_\lambda}^{i}[\alpha_k\ q_k\ \delta_{ek}]} \tag{7.37b}$$

Defuzzify the $C_{Lmn,\underline{\alpha}_\lambda}([\alpha_k\ q_k\ \delta_{ek}])$ and $C_{Lmn,\overline{\alpha}_\lambda}([\alpha_k\ q_k\ \delta_{ek}])$ for $\underline{\alpha}_1, \underline{\alpha}_2, \ldots, \underline{\alpha}_\rho$ and $\overline{\alpha}_1, \overline{\alpha}_2, \ldots, \overline{\alpha}_\rho$, and generate the IT3 FLS as (7.38)

$$C_{Lmn\text{-IT3}}([\alpha_k\ q_k\ \delta_{ek}]) = \frac{\sum_{\lambda=1}^{\rho}(\underline{\alpha}_\lambda C_{Lmn,\underline{\alpha}_\lambda}([\alpha_k\ q_k\ \delta_{ek}]) + \overline{\alpha}_\lambda C_{Lmn,\overline{\alpha}_\lambda}([\alpha_k\ q_k\ \delta_{ek}]))}{\sum_{\lambda=1}^{\rho}(\underline{\alpha}_\lambda + \overline{\alpha}_\lambda)} \tag{7.38}$$

The $m, n$ tuning parameters are optimized for mean square error (MSE) minimization and are found by utilizing the algorithms offered in (7.39).

$$\text{MSE}_L = \frac{1}{N}\sum_{k=1}^{N}(C_{\text{Lmn-IT3}}([\alpha_k \quad q_k \quad \delta_{ek}]) - C_{Lk})^2 \tag{7.39}$$

### *7.3.3 IT3 T-S Fuzzy Model for Pitching Moment Coefficient*

The IT3 T-S fuzzy model for the pitching moment coefficient is constructed using the data pairs $([\alpha_k, q_k, \delta_{ek}], C_{Mk})$ for $k = 1, 2, \ldots, N$. Its input variables are identical to those of the drag coefficient model, while the output variable is the pitching moment coefficient $C_M$. The mathematical formulation presented in Sect. 7.3.1 can be applied here with only a change in subscripts. Similar to the drag model, the pitching moment coefficient is represented through a hyperplane regression function, expressed as

$$\hat{C}_{M_k} = f^i\left([\alpha_k\ q_k\ \delta_{e_k}], \boldsymbol{\omega}_{Mi}\right) = a_{M\alpha}^i\alpha_k + a_{Mq}^i q_k + a_{M\delta_e}^i\delta_{e_k} + a_{M0}^i, \tag{7.40}$$

where, $\boldsymbol{\omega}_{M_i} = \left[a_{M\alpha}^i,\ a_{Mq}^i,\ a_{M\delta_e}^i,\ a_{M0}^i\right]$ is the hyperplane model coefficient vector.

The upper and lower hyperplanes are obtained as:

$$\overline{\boldsymbol{\omega}}_{M_i} = \left[\mathbf{X}^T\overline{\mathbf{P}}_{M_i}[\mathbf{X}]\right]^{-1}\mathbf{X}^T\overline{\mathbf{P}}_{M_i}\mathbf{C}_M, \tag{7.41}$$

$$\underline{\boldsymbol{\omega}}_{M_i} = \left[\mathbf{X}^T\underline{\mathbf{P}}_{M_i}\mathbf{X}\right]^{-1}\mathbf{X}^T\underline{\mathbf{P}}_{M_i}\mathbf{C}_M, \tag{7.42}$$

where partition matrices $\overline{\mathbf{P}}_{M_i}$ and $\underline{\mathbf{P}}_{M_i}$ are

$$\overline{\mathbf{P}}_{M_i} = \begin{bmatrix} \overline{u}_{Mi1} & 0 & \cdots & 0 \\ 0 & \overline{u}_{Mi2} & \cdots & 0 \\ \vdots & \vdots & \ddots & \vdots \\ 0 & 0 & \cdots & \overline{u}_{MiN} \end{bmatrix}, \quad \underline{\mathbf{P}}_{M_i} = \begin{bmatrix} \underline{u}_{Mi1} & 0 & \cdots & 0 \\ 0 & \underline{u}_{Mi2} & \cdots & 0 \\ \vdots & \vdots & \ddots & \vdots \\ 0 & 0 & \cdots & \underline{u}_{MiN} \end{bmatrix} \in R^{N\times N}.$$

Like the drag coefficient model, the hyperplane-shaped Gaussian MF is utilized for each input parameter. The upper and lower membership degrees for each input parameter are as follows:

For first input parameter $\alpha_k$:

$$\overline{\mu}_{\tilde{M}_i}(\alpha_k) = \exp\left(-\eta_M \cdot \frac{d_{Mi\alpha_k}(\overline{\boldsymbol{\omega}}_{M_i})}{\max\left(\{d_{Mi\alpha_k}(\overline{\boldsymbol{\omega}}_{M_i})\},\ i = 1,\ldots,c\right)}\right), \tag{7.43a}$$

$$\underline{\mu}_{\tilde{M}_i}(\alpha_k) = \exp\left(-\eta_M \cdot \frac{d_{Mi_{\alpha_k}}(\underline{\boldsymbol{\omega}}_{Mi})}{\max(\{d_{Mi_{\alpha_k}}(\underline{\boldsymbol{\omega}}_{Mi}), i = 1,\ldots,c\})}\right), \tag{7.43b}$$

where, $d_{Mi\alpha_k}(\bar{\boldsymbol{\omega}}_{Mi})$ represent the distance of $\alpha_k$ from $i$th upper and $d_{Mi\alpha_k}(\underline{\boldsymbol{\omega}}_{Mi})$ is the distance of $\alpha_k$ from $i$th lower hyper planes. The parameter $\eta_M$ is a constant, which is used for tuning of upper and lower membership grades $\bar{\mu}_{\tilde{M}_i}(\alpha_k)$ and $\underline{\mu}_{\tilde{M}_i}(\alpha_k)$. These parameters regulate the fuzziness of the membership grades and are referred to as the antecedent membership function parameters. The parameters $d_{Mi\alpha_k}(\bar{\boldsymbol{\omega}}_{Mi})$ and $d_{Mi\alpha_k}(\underline{\boldsymbol{\omega}}_{Mi})$ are obtained as follows:

$$d_{Mi\alpha_k}(\bar{\boldsymbol{\omega}}_{Mi}) = \frac{|\alpha_k \cdot \bar{\boldsymbol{\omega}}_{Mi}|}{\|\bar{\boldsymbol{\omega}}_{Mi}\|} \quad \text{and} \quad d_{Mi\alpha_k}(\underline{\boldsymbol{\omega}}_{Mi}) = \frac{|\alpha_k \cdot \underline{\boldsymbol{\omega}}_{Mi}|}{\|\underline{\boldsymbol{\omega}}_{Mi}\|}$$

For second input parameter $q_k$:

$$\bar{\mu}_{\tilde{M}_i}(q_k) = \exp\left(-\eta_M \cdot \frac{d_{Miq_k}(\bar{\boldsymbol{\omega}}_{Mi})}{\max(\{d_{Miq_k}(\bar{\boldsymbol{\omega}}_{Mi}), i = 1,\ldots,c\})}\right), \tag{7.44a}$$

$$\underline{\mu}_{\tilde{M}_i}(q_k) = \exp\left(-\eta_M \cdot \frac{d_{Miq_k}(\underline{\boldsymbol{\omega}}_{Mi})}{\max(\{d_{Miq_k}(\underline{\boldsymbol{\omega}}_{Mi}), i = 1,\ldots,c\})}\right), \tag{7.44b}$$

where, $d_{Miq_k}(\bar{\omega}_{Mi})$ is the distance of $q_k$ from $i$th upper and $d_{Miq_k}(\underline{\omega}_{Mi})$ is the distance of $q_k$ from $i$th lower hyper planes

$$d_{Miq_k}(\bar{\boldsymbol{\omega}}_{Mi}) = \frac{|q_k \cdot \bar{\boldsymbol{\omega}}_{Mi}|}{\|\bar{\boldsymbol{\omega}}_{Mi}\|} \quad \text{and} \quad d_{Miq_k}(\underline{\boldsymbol{\omega}}_{Mi}) = \frac{|q_k \cdot \underline{\boldsymbol{\omega}}_{Mi}|}{\|\underline{\boldsymbol{\omega}}_{Mi}\|}$$

For third input parameter $\delta_{ek}$:

$$\bar{\mu}_{\tilde{M}_i}(\delta_{ek}) = \exp\left(-\eta_M \cdot \frac{d_{Mi\delta_{ek}}(\bar{\boldsymbol{\omega}}_{Mi})}{\max(\{d_{Mi\delta_{ek}}(\bar{\boldsymbol{\omega}}_{Mi}), i = 1, \ldots, c\})}\right), \tag{7.45a}$$

$$\underline{\mu}_{\tilde{M}_i}(\delta_{ek}) = \exp\left(-\eta_M \cdot \frac{d_{Mi\delta_{ek}}(\underline{\boldsymbol{\omega}}_{Mi})}{\max(\{d_{Mi\delta_{ek}}(\underline{\boldsymbol{\omega}}_{Mi}), i = 1, \ldots, c\})}\right), \tag{7.45b}$$

where, $d_{Mi\delta_{ek}}(\bar{\omega}_{Mi})$ is the distance of $\delta_{ek}$ from $i$th upper and $d_{Mi\delta_{ek}}(\underline{\omega}_{Mi})$ is the distance of $\delta_{ek}$ from $i$th lower hyper planes

$$d_{Mi_{\delta_{ek}}}(\bar{\boldsymbol{\omega}}_{Mi}) = \frac{|\delta_{ek} \cdot \bar{\boldsymbol{\omega}}_{Mi}|}{\|\bar{\boldsymbol{\omega}}_{Mi}\|} \quad \text{and} \quad d_{Mi_{\delta_{ek}}}(\underline{\boldsymbol{\omega}}_{Mi}) = \frac{|\delta_{ek} \cdot \underline{\boldsymbol{\omega}}_{Mi}|}{\|\underline{\boldsymbol{\omega}}_{Mi}\|}$$

An IT2 T-S FLS for modeling of the Lift parameter is defined by the following IF-THEN fuzzy rules ($i$th rule)

$$R_M^i : \text{ IF } \alpha \text{ is } \tilde{M}_\alpha^i \text{ and } q \text{ is } \tilde{M}_q^i \text{ and } \delta_e \text{ is } \tilde{M}_{\delta_e}^i$$

THEN

$$C_M^i(\alpha, q, \delta_e) = C_{M\alpha}^i \, \alpha + C_{Mq}^i \, q + C_{M\delta_e}^i \, \delta_e + C_{M0}^i,$$

where $\tilde{M}_\alpha^i$, $\tilde{M}_q^i$, $\tilde{M}_{\delta_e}^i$ are fuzzy model terms for the drag coefficient model related to each input parameter.

The interval-type consequent parameters $\left(C_{M\alpha}^i, C_{Mq}^i, C_{M\delta_e}^i, C_{M0}^i\right)$ of the IT2 T-S FLS are derived from the upper and lower hyperplane coefficients $\bar{\omega}_{Mi}$ and $\underline{\omega}_{Mi}$. Similar to the drag coefficient model, the consequent of rule $R_M^i$ is given as $C_M^i = [C_{Ml}^i, C_{Mr}^i]$, obtained from the lower and upper hyperplane coefficient vectors.

The base of SMF (triangular type) for each input equals $\left(\overline{\mu}_{\tilde{M}}(\alpha) - \underline{\mu}_{\tilde{M}}(\alpha)\right)$, $\left(\overline{\mu}_{\tilde{M}}(q) - \underline{\mu}_{\tilde{M}}(q)\right)$, $\left(\overline{\mu}_{\tilde{M}}(\delta_e) - \underline{\mu}_{\tilde{M}}(\delta_e)\right)$ and its apex location is obtained as [14].

$$\text{apex}(\alpha) = \underline{\mu}_{\tilde{M}}(\alpha) + w_M \left[\overline{\mu}_{\tilde{M}}(\alpha) - \underline{\mu}_{\tilde{M}}(\alpha)\right], \tag{7.46a}$$

$$\text{apex}(q) = \underline{\mu}_{\tilde{M}}(q) + w_M \left[\overline{\mu}_{\tilde{M}}(q) - \underline{\mu}_{\tilde{M}}(q)\right], \tag{7.46b}$$

$$\text{apex}(\delta_e) = \underline{\mu}_{\tilde{M}}(\delta_e) + w_M \left[\overline{\mu}_{\tilde{M}}(\delta_e) - \underline{\mu}_{\tilde{M}}(\delta_e)\right], \tag{7.46c}$$

where $w_M \in [0, 1]$ signifies the design parameter. The $\alpha$- cuts of triangle SMFs for first input parameter is given as.

$$\begin{cases} \tilde{M}_\alpha(\alpha) = \left[a_{\tilde{M}\alpha}(\alpha), b_{\tilde{M}\alpha}(\alpha)\right] \\ a_{\tilde{M}\alpha}(\alpha) = \underline{\mu}_{\tilde{M}}(\alpha) + w_M\left[\overline{\mu}_{\tilde{M}}(\alpha) - \underline{\mu}_{\tilde{M}}(\alpha)\right]\alpha \\ b_{\tilde{M}\alpha}(\alpha) = \overline{\mu}_{\tilde{M}}(\alpha) + (1 - w_M)\left[\overline{\mu}_{\tilde{M}}(\alpha) - \underline{\mu}_{\tilde{M}}(\alpha)\right]\alpha \end{cases} \tag{7.47a}$$

Similarly, $\alpha$ cuts of triangle SMFs for the second and third input parameters $q$ and $\delta_e$ are given as (7.47b) and (7.47c).

$$\begin{cases} \tilde{M}_\alpha(q) = \left[a_{\tilde{L}\alpha}(q), b_{\tilde{M}\alpha}(q)\right] \\ a_{\tilde{M}\alpha}(q) = \underline{\mu}_{\tilde{M}}(q) + w_M\left[\overline{\mu}_{\tilde{M}}(q) - \underline{\mu}_{\tilde{M}}(q)\right]\alpha \\ b_{\tilde{M}\alpha}(q) = \overline{\mu}_{\tilde{M}}(q) + (1 - w_M)\left[\overline{\mu}_{\tilde{M}}(q) - \underline{\mu}_{\tilde{M}}(q)\right]\alpha \end{cases} \tag{7.47b}$$

$$\begin{cases} \tilde{M}_\alpha(\delta_e) = \left[a_{\tilde{M}\alpha}(\delta_e), b_{\tilde{M}\alpha}(\delta_e)\right] \\ a_{\tilde{M}\alpha}(\delta_e) = \underline{\mu}_{\tilde{M}}(\delta_e) + w_M\left[\overline{\mu}_{\tilde{M}}(\delta_e) - \underline{\mu}_{\tilde{M}}(\delta_e)\right]\alpha \\ b_{\tilde{M}\alpha}(\delta_e) = \overline{\mu}_{\tilde{M}}(\delta_e) + (1 - w_M)\left[\overline{\mu}_{\tilde{M}}(\delta_e) - \underline{\mu}_{\tilde{M}}(\delta_e)\right]\alpha \end{cases} \tag{7.47c}$$

The $\alpha$-cut decomposition for each rule-antecedent of type-1 vertical slice (when $[\alpha\ q\ \delta_e] = [\alpha', q', \delta_e']$) $\tilde{M}^i_\alpha(\alpha')$, $\tilde{M}^i_\alpha(q')$ and $\tilde{M}^i_\alpha(\delta_e')$ is obtained as

$$\tilde{M}^i_\alpha(\alpha') = \sup_{\forall\alpha\in[0,1]} \alpha/\left[a^i_{\tilde{M}\alpha}(\alpha'), b^i_{\tilde{M}\alpha}(\alpha')\right] \tag{7.48a}$$

$$\tilde{M}^i_\alpha(q') = \sup_{\forall\alpha\in[0,1]} \alpha/\left[a^i_{\tilde{M}\alpha}(q'), b^i_{\tilde{M}\alpha}(q')\right] \tag{7.48b}$$

$$\tilde{M}^i_\alpha(\delta_e') = \sup_{\forall\alpha\in[0,1]} \alpha/\left[a^i_{\tilde{M}\alpha}(\delta_e'), b^i_{\tilde{M}\alpha}(\delta_e')\right] \tag{7.48c}$$

The $\alpha$ level firing interval, $\tilde{M}^i_\alpha[\alpha'\ q'\ \delta_e']$ is as

$$\begin{aligned} \tilde{M}^i_\alpha[\alpha'\ q'\ \delta_e'] &= [\underline{l}^i_\alpha[\alpha'\ q'\ \delta_e'], \overline{m}^i_\alpha[\alpha'\ q'\ \delta_e']] \\ \underline{m}^i_\alpha[\alpha'\ q'\ \delta_e'] &= \min(a^i_{\tilde{M}\alpha}(\alpha'), a^i_{\tilde{M}\alpha}(q'), a^i_{\tilde{M}\alpha}(\delta_e')) \\ \overline{m}^i_\alpha[\alpha'\ q'\ \delta_e'] &= \min(b^i_{\tilde{M}\alpha}(\alpha'), b^i_{\tilde{M}\alpha}(q'), b^i_{\tilde{M}\alpha}(\delta_e')) \end{aligned} \tag{7.49}$$

The lower SMF ($\underline{\alpha}$) and upper SMF ($\overline{\alpha}$) cut points are represented as $\underline{\alpha}/[a_{\tilde{M}\underline{\alpha}}(\alpha'), b_{\tilde{M}\underline{\alpha}}(\alpha')]$, $\underline{\alpha}/[a_{\tilde{M}\underline{\alpha}}(q'), b_{\tilde{M}\underline{\alpha}}(q')]$, $\underline{\alpha}/[a_{\tilde{M}\underline{\alpha}}(\delta_e'), b_{\tilde{M}\underline{\alpha}}(\delta_e')]$ and $\overline{\alpha}/[a_{\tilde{M}\overline{\alpha}}(\alpha'), b_{\tilde{M}\overline{\alpha}}(\alpha')]$, $\overline{\alpha}/[a_{\tilde{M}\overline{\alpha}}(q'), b_{\tilde{M}\overline{\alpha}}(q')]$, $\overline{\alpha}/[a_{\tilde{M}\overline{\alpha}}(\delta_e'), b_{\tilde{M}\overline{\alpha}}(\delta_e')]$, respectively. Like the drag coefficient model, these cut points are estimated for discretized $\underline{\alpha}_\lambda$ and $\overline{\alpha}_\lambda$ levels similar to (7.20a)–(7.20c) and (7.21a)–(7.21c) for $\lambda = 1, \ldots, \rho$.

Further the firing interval $\underline{\tilde{M}}^{i}_{\underline{\alpha}_\lambda}[\alpha_k \quad q_k \quad \delta_{e_k}]$ and $\tilde{M}_{\overline{\alpha}_\lambda}[\alpha_k \quad q_k \quad \delta_{e_k}]$ for each $\underline{\alpha}_\lambda$ and $\overline{\alpha}_\lambda$ level corresponding to $i$th rule are as

$$\begin{cases} \widetilde{M}^{i}_{\underline{\alpha}_\lambda}[\alpha_k, q_k, \delta_{ek}] \equiv \left[\underline{m}^{i}_{\underline{\alpha}_\lambda}[\alpha_k, q_k, \delta_{ek}], \overline{m}^{i}_{\underline{\alpha}_\lambda}[\alpha_k, q_k, \delta_{ek}]\right], \\ \underline{m}^{i}_{\underline{\alpha}_\lambda}[\alpha_k, q_k, \delta_{ek}] = \min\left(a^{i}_{\widetilde{M}\underline{\alpha}_\lambda}(\alpha_k), a^{i}_{\widetilde{M}\underline{\alpha}_\lambda}(q_k), a^{i}_{\widetilde{M}\underline{\alpha}_\lambda}(\delta_{ek})\right), \\ \overline{m}^{i}_{\underline{\alpha}_\lambda}[\alpha_k, q_k, \delta_{ek}] = \min\left(b^{i}_{\widetilde{M}_{\underline{\alpha}_\lambda}}(\alpha_k), b^{i}_{\widetilde{M}\underline{\alpha}_\lambda}(q_k), b^{i}_{\widetilde{M}\underline{\alpha}_\lambda}(\delta_{ek})\right) \end{cases} \tag{7.50a}$$

$$\begin{cases} \widetilde{M}^{i}_{\overline{\alpha}_\lambda}[\alpha_k, q_k, \delta_{ek}] \equiv \left[\underline{m}^{i}_{\overline{\alpha}_\lambda}[\alpha_k, q_k, \delta_{ek}], \overline{m}^{i}_{\overline{\alpha}_\lambda}[\alpha_k, q_k, \delta_{ek}]\right], \\ \underline{m}^{i}_{\overline{\alpha}_\lambda}[\alpha_k, q_k, \delta_{ek}] = \min\left(a^{i}_{\widetilde{M}\overline{\alpha}_\lambda}(\alpha_k), a^{i}_{\widetilde{M}\overline{\alpha}_\lambda}(q_k), a^{i}_{\widetilde{M}\overline{\alpha}_\lambda}(\delta_{ek})\right), \\ \overline{m}^{i}_{\overline{\alpha}_\lambda}[\alpha_k, q_k, \delta_{ek}] = \min\left(b^{i}_{\widetilde{M}\overline{\alpha}_\lambda}(\alpha_k), b^{i}_{\widetilde{M}\overline{\alpha}_\lambda}(q_k), b^{i}_{\widetilde{M}\overline{\alpha}_\lambda}(\delta_{ek})\right) \end{cases} \tag{7.50b}$$

In the next step $\underline{\alpha}_\lambda$ and $\overline{\alpha}_\lambda$ level $m$-$n$ outputs $C_{Mmn,\underline{\alpha}_\lambda}([\alpha_k \; q_k \; \delta_{ek}])$ and $C_{Mmn,\overline{\alpha}_\lambda}([\alpha_k \; q_k \; \delta_{ek}])$ at each $\underline{\alpha}_1, \underline{\alpha}_2, \ldots, \underline{\alpha}_\rho$ and $\overline{\alpha}_1, \overline{\alpha}_2, \ldots, \overline{\alpha}_\rho$ as,

$$C_{Mmn,\underline{\alpha}_\lambda}([\alpha_k \; q_k \; \delta_{ek}]) = m\frac{\sum_{i=1}^{c} C^{i}_{Ml}\underline{m}^{i}_{\underline{\alpha}_\lambda}[\alpha_k \; q_k \; \delta_{ek}]}{\sum_{i=1}^{c} \underline{m}^{i}_{\underline{\alpha}_\lambda}[\alpha_k \; q_k \; \delta_{ek}]} + n\frac{\sum_{i=1}^{c} C^{i}_{Mr}\overline{m}^{i}_{\underline{\alpha}_\lambda}[\alpha_k \; q_k \; \delta_{ek}]}{\sum_{i=1}^{c} \overline{m}^{i}_{\underline{\alpha}_\lambda}[\alpha_k \; q_k \; \delta_{ek}]} \tag{7.51a}$$

$$C_{Mmn,\bar{\alpha}_\lambda}([\alpha_k \; q_k \; \delta_{ek}]) = m\frac{\sum_{i=1}^{c} C^{i}_{Ml}\underline{m}^{i}_{\bar{\alpha}_\lambda}[\alpha_k \; q_k \; \delta_{ek}]}{\sum_{i=1}^{c} \underline{m}^{i}_{\overline{\alpha}_\lambda}[\alpha_k \; q_k \; \delta_{ek}]} + n\frac{\sum_{i=1}^{c} C^{i}_{Mr}\overline{m}^{i}_{\overline{\alpha}_\lambda}[\alpha_k \; q_k \; \delta_{ek}]}{\sum_{i=1}^{c} \overline{m}^{i}_{\overline{\alpha}_\lambda}[\alpha_k \; q_k \; \delta_{ek}]} \tag{7.51b}$$

Defuzzify the $C_{Mmn,\underline{\alpha}_\lambda}([\alpha_k \; q_k \; \delta_{ek}])$ and $C_{Mmn,\overline{\alpha}_\lambda}([\alpha_k \; q_k \; \delta_{ek}])$ for $\underline{\alpha}_1, \underline{\alpha}_2, \ldots, \underline{\alpha}_\rho$ and $\overline{\alpha}_1, \overline{\alpha}_2, \ldots, \overline{\alpha}_\rho$, and generate the IT3 FLS as (7.52)

$$C_{Mmn\text{-IT3}}([\alpha_k \; q_k \; \delta_{ek}]) = \frac{\sum_{\lambda=1}^{\rho}(\underline{\alpha}_\lambda C_{Mmn,\underline{\alpha}_\lambda}([\alpha_k \; q_k \; \delta_{ek}]) + \overline{\alpha}_\lambda C_{Mmn,\overline{\alpha}_\lambda}([\alpha_k \; q_k \; \delta_{ek}]))}{\sum_{\lambda=1}^{\rho}(\underline{\alpha}_\lambda + \overline{\alpha}_\lambda)} \tag{7.52}$$

For minimization of MSE, the values of $m$ and $n$ tuning parameters are optimized as per (7.53).

$$\text{MSE}_M = \frac{1}{N}\sum_{k=1}^{N}(C_{\text{Mmn-IT3}}([\alpha_k \quad q_k \quad \delta_{ek}]) - C_{Mk})^2 \tag{7.53}$$

## 7.4 Simulation Results: IT3 T-S Fuzzy Model for Longitudinal Aerodynamic Coefficient of ATTAS Aircraft

### *7.4.1 Test Flight Data Set*

To demonstrate the modeling, flight data were obtained from the ATTAS aircraft of DLR German Aerospace Center [4, 39]. The aircraft operated under quasi-steady stall conditions at an altitude of 16,000 ft. The dataset spans 200 seconds (5000 samples) with a sampling interval of 0.04s. Two pilot-executed maneuvers are considered: the first over 120s and the second over 80s [4]. The IT3 T-S fuzzy system is employed to capture aircraft dynamics without explicitly referring to the underlying stall physics. Out of the complete dataset (5000 samples, 200s), 4000 samples (160s) are used for modeling and 1000 samples (40s) for testing. To assess robustness, the training and testing datasets are further augmented with different levels of noise and various uncertainty scenarios. Finally, the performance of the developed IT3 T-S fuzzy aerodynamic model is compared against other state-of-the-art modeling techniques.

### *7.4.2 Performance Analysis of Drag Coefficient IT3 T-S Fuzzy Model*

The performance of the IT3 T-S fuzzy model for the drag coefficient is assessed using quasi-steady stall maneuvering flight data, where $C_D$ is represented as a nonlinear function of angle of attack ($\alpha$), pitch rate ($q$), and elevator deflection ($\delta_e$) as,

$$C_D = f_{nonlinear}(\alpha, q, \delta_e) \tag{7.54}$$

The training dataset consists of 5000 samples (200s), while the testing dataset contains 1000 samples (40s). The IT3 T-S fuzzy system is configured with the following parameters:

- Number of fuzzy rules: $c = 7$,
- Fuzziness parameters: $m_1 = 1.41,\ m_2 = 5.2$,
- MF tuning parameter: $\eta_D = 95$,
- Secondary MF tuning parameter: $w_D = 0.5$.

The $\alpha$-plane values are defined as

$$[\alpha_1, \alpha_2, \alpha_3, \alpha_4, \alpha_5, \alpha_6] = [0, 0.2, 0.4, 0.6, 0.8, 1.0],$$

with fuzziness $\Delta = 2$. The tuning parameters are set to $m = 0.55$ and $n = 0.45$.

**Case A**: Three versions of the training dataset are created by adding zero-mean Gaussian noise with standard deviations of 0.1, 0.2, and 0.3 to the clean data. For testing, one clean dataset (1000 samples, 40s) and two noise-augmented sets (with standard deviations of 0.1 and 0.3) are used.

**Case B**: From both training and testing datasets, 200 samples of the angle of attack parameter ($\alpha$) are randomly replaced with zeros, simulating the effect of missing data combined with noise.

**Case C**: In this case, 200 samples of the angle of attack ($\alpha$) from the training and testing datasets are substituted with extreme values, representing invariability in data along with noise.

**Evaluation**: The drag coefficient estimated by the IT3 T-S fuzzy model, $\hat{C}_D$, is compared against the actual coefficient $C_D$ for both training and testing. Figure 7.1a and b illustrate the model's prediction performance under noise-free and noisy conditions (noise SD = 0.1). The results show that the IT3 T-S fuzzy model effectively handles uncertainties arising from different noise levels and data irregularities.

### *7.4.3 Performance Analysis of Lift Coefficient IT3 T-S Fuzzy Model*

The performance of the IT3 T-S fuzzy model for the lift coefficient is evaluated using quasi-steady stall maneuvering flight data, where $C_L$ is modeled as a nonlinear function of angle of attack ($\alpha$), pitch rate ($q$), and elevator deflection ($\delta_e$) as,

$$C_L = f_{nonlinear}(\alpha, q, \delta_e). \tag{7.55}$$

The dataset consists of 5000 samples (200s) for training, while 1000 samples (40s) are reserved for testing. The IT3 T-S fuzzy model is configured with the following parameters:

- Number of fuzzy rules: $c = 7$,
- Fuzziness parameters: $m_1 = 1.75, \ m_2 = 5.7$,
- MF tuning parameter: $\eta_L = 93$,
- Secondary MF tuning parameter: $w_L = 0.5$.

The $\alpha$-plane parameters are defined as

$$[\alpha_1, \alpha_2, \alpha_3, \alpha_4, \alpha_5, \alpha_6] = [0, 0.2, 0.4, 0.6, 0.8, 1.0],$$

with a fuzziness level of $\Delta = 2$. The tuning parameters are set as $m = 0.57$ and $n = 0.43$.

To assess the performance of the developed IT3 T-S fuzzy model for the lift coefficient, the estimated values ($\hat{C}_L$) are compared with the actual lift coefficient ($C_L$) using both training and testing datasets. Figure 7.2a and b illustrate the prediction

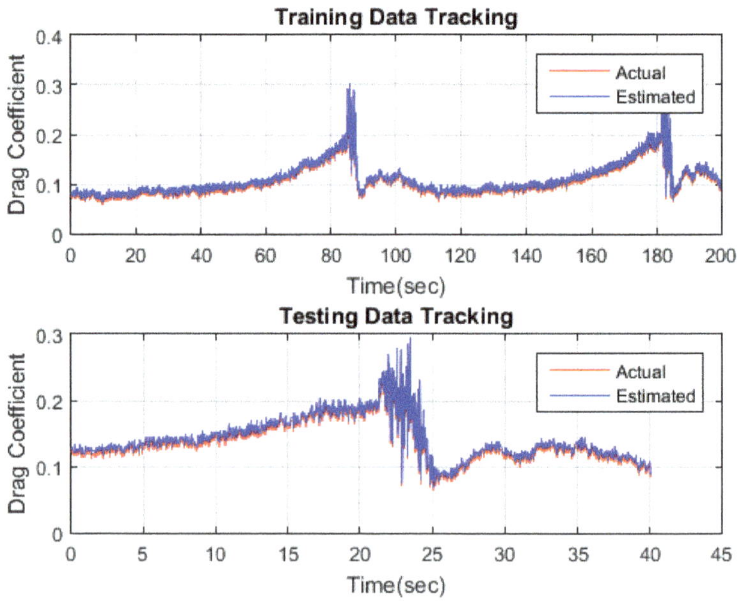


(a)

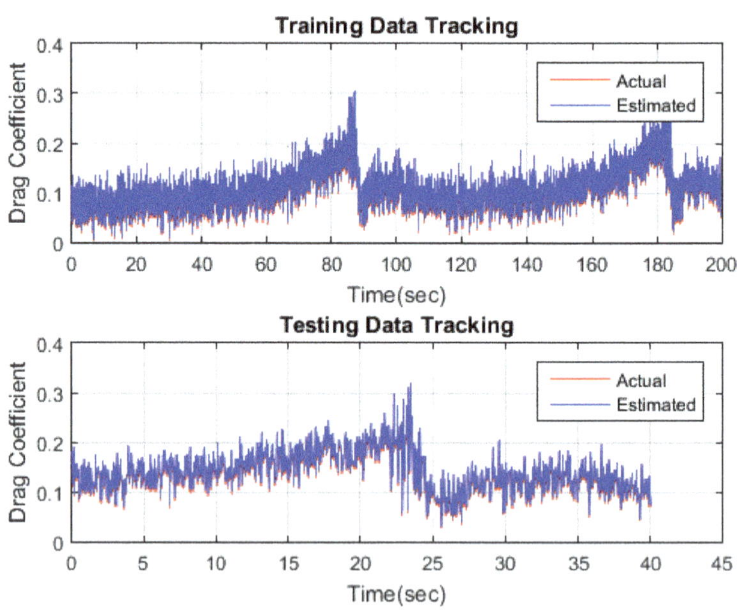


(b)

**Fig. 7.1** Performance of the IT3 fuzzy model for drag coefficient prediction: **a** noise-free case, **b** case with added noise (0.1 SD in both training and testing)

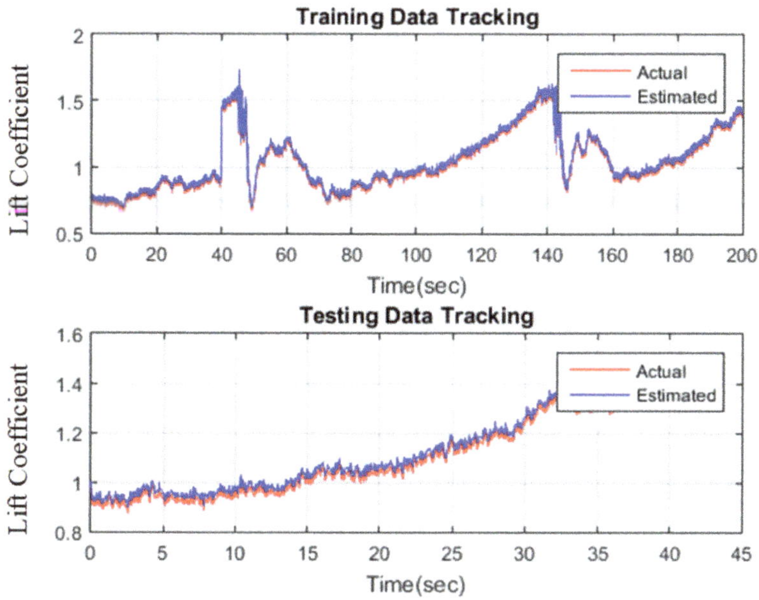


(a)

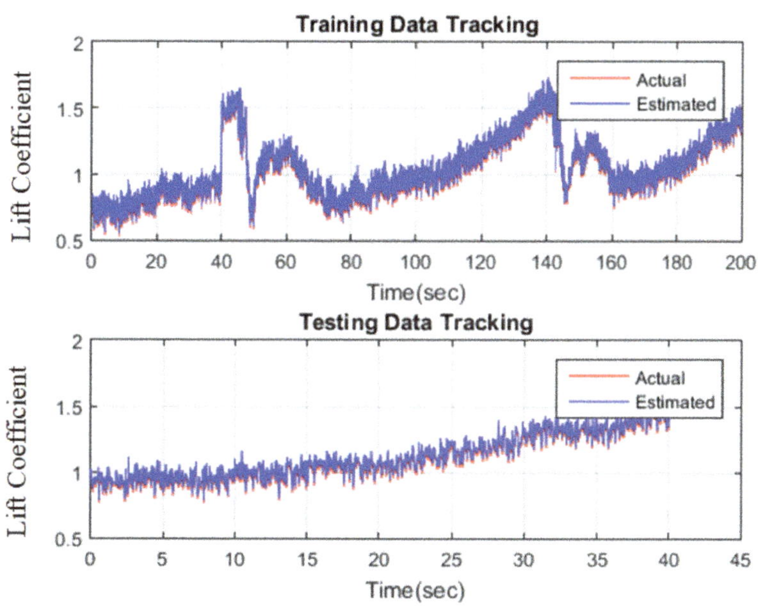


(b)

**Fig. 7.2** Performance of the IT3 fuzzy model for lift coefficient prediction: **a** noise-free case, **b** case with added noise (0.1 SD in both training and testing)

results for two cases: without noise and with noise (0.1 SD added to both training and testing data).

### 7.4.4 *Performance Analysis of Pitching Moment Coefficient IT3 T-S Fuzzy Model*

The flying data from quasi-steady stall maneuvering is used to model the pitching moment coefficient. It's nonlinear function is shown as

$$C_M = f_{nonlinear}(\alpha, q, \delta_e) \tag{7.56}$$

The training dataset consists of 5000 samples (200s), while 1000 samples (40s) are used for testing. The IT3 T-S fuzzy model is configured with the following parameters:

- Number of fuzzy rules: $c = 7$,
- Fuzziness parameters: $m_1 = 2.15,\ \ m_2 = 6.3$,
- MF tuning parameter: $\eta_M = 81$,
- Secondary MF tuning parameter: $w_M = 0.5$.

The $\alpha$-plane parameters are defined as

$$[\alpha_1, \alpha_2, \alpha_3, \alpha_4, \alpha_5, \alpha_6] = [0, 0.2, 0.4, 0.6, 0.8, 1.0],$$

with an $\alpha$-plane fuzziness level of $\Delta = 2$. The tuning parameters are set as $m = 0.51$ and $n = 0.49$.

Time histories of the estimated pitching moment coefficient parameter ($\hat{C}_M$) and the actual pitching moment coefficient parameter ($C_M$) are compared. The prediction performance of the developed model is displayed in Fig. 7.3a and b for both the noise-related and noise-free scenarios.

## 7.5 Summary

This chapter presents a method for simulating quasi-stationary stall behavior in longitudinal aerodynamics at high angles of attack using the IT3 T-S fuzzy logic framework. Building on the theoretical foundation established by Singh et al. [32], the approach is tailored for aerodynamic applications. The model captures lift, drag, and pitching moment coefficients through IT3 T-S fuzzy logic, employing pre-processed flight data from the ATTAS aircraft for validation. To evaluate robustness, datasets with different noise levels and uncertainties–such as missing or invariant data–are introduced. Results show that the proposed IT3 T-S fuzzy system effectively manages

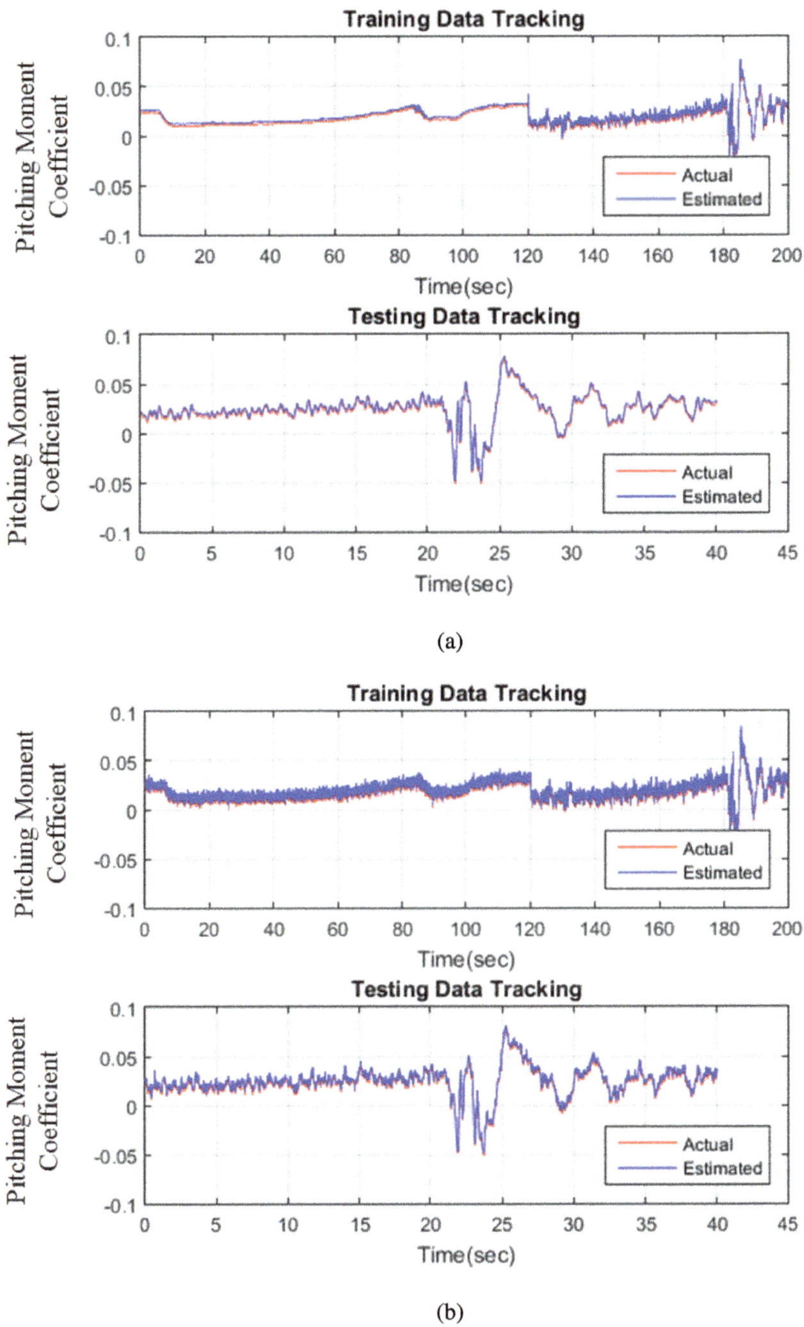


**Fig. 7.3** Performance of the IT3 fuzzy model for pitching moment coefficient prediction: **a** noise-free case, **b** case with added noise (0.1 SD in both training and testing)

these uncertainties and noise, offering reliable aerodynamic predictions. When compared with advanced modeling techniques, the IT3 T-S fuzzy model demonstrates superior performance, particularly under corrupted or uncertain data conditions.

## References

1. P. G. Hamel and R.V. Jategaonkar, "Evolution of Flight Vehicle System Identification," Journal of Aircraft, Vol. 33, No. 1, Jan–Feb 1996.
2. M. Tischler, and R. Remple, "Aircraft and Rotorcraft System Identification," AIAA, Reston, VA, 2006.
3. P. G. Hamel, R.V. Jategaonkar, and K. F. Doherr, "Identification of the Aerodynamic Model of the DLR Research Aircraft ATTAS from Flight Test Data," DLR Rept. DLR-FB 90-40, Braunschweig, Germany, 1990.
4. R. V. Jategaonkar, "Flight Vehicle System Identification A Time Domain Methodology," American Institute of Aeronautics and Astronautics Inc., Reston, Virginia 2006.
5. O. Nelles, "Nonlinear System Identification from Classical Approaches to Neural Networks and Fuzzy Models", Springer-Verlag Berlin Heidelberg 2001.
6. H. Verma and N. Peyada, "Parameter Estimation of Unstable Aircraft using Extreme Learning Machine", Defence Science Journal, 67(6), 603–611, 2017. https://doi.org/10.14429/dsj.67.11401.
7. Jay M.Brandon and Eugene A.Morelli, "Nonlinear Aerodynamic Modeling From Flight Data Using Advanced Piloted Maneuvers and Fuzzy Logic", "NASA Center for Aerospace Information Langley Research Center Hampton, Virginia", NASA/TM-2012-217778 October 2012.
8. Ray C. Chang, "Fuzzy Logic-Based Aerodynamic Modeling with Continuous Differentiability", "Hindwai Publishing Corporation , Mathematical Problems in En-ginering", "Volume 2013, Article ID 609769".
9. Qing Wang, Qian Weiqi, He Kaifeng, "Unsteady aerodynamic modeling at high angles of attack using support vector machines", ""CSAA Chinese Journal of Aeronautics (2015), 28(3): 659–668.
10. .R. Chambers, "High-Angle-of-Attack Technology: Progress and Challenges", NASA CP-3149, vol. 1, May 1992, pp. 1–22.
11. Min-You Chen and D.A Linkens, "A Systematic Method for Fuzzy Modeling from Numerical Data", "0-7803-7087-2/01 IEEE 2001.
12. Takagi, T. and Sugeno M,."Fuzzy identification of systems and its application to modeling and control," IEEE Transactions on Syst,Man ,Cyber Vol SMC-15 Jan/Feb 1985: 116–132.
13. J. M. Mendel, R. I. John and F. Liu, "Interval Type-2 Fuzzy Logic Systems Made Simple," in IEEE Transactions on Fuzzy Systems, vol. 14, no. 6, pp. 808–821, Dec. 2006, https://doi.org/10.1109/TFUZZ.2006.879986.
14. J. M. Mendel, "General Type-2 Fuzzy Logic Systems Made Simple: A Tutorial," in IEEE Transactions on Fuzzy Systems, vol. 22, no. 5, pp. 1162–1182, Oct. 2014, https://doi.org/10.1109/TFUZZ.2013.2286414.
15. J. M. Mendel, "Comparing the Performance Potentials of Interval and General Type-2 Rule-Based Fuzzy Systems in Terms of Sculpting the State Space," in IEEE Transactions on Fuzzy Systems, vol. 27, no. 1, pp. 58–71, Jan. 2019, https://doi.org/10.1109/TFUZZ.2018.2856184.
16. A. Amirkhani, M. Shirzadeh and M. Molaie, "An Indirect Type-2 Fuzzy Neural Network Optimized by the Grasshopper Algorithm for Vehicle ABS Controller," in IEEE Access, vol. 10, pp. 58736–58751, 2022, https://doi.org/10.1109/ACCESS.2022.3179700.
17. Amirkhani, A, Shirzadeh, M, Kumbasar, T, Mashadi, B. A framework for designing cognitive trajectory controllers using genetically evolved interval type-2 fuzzy cognitive maps. Int J Intell Syst. 2022; 37: 305–335. https://doi.org/10.1002/int.22626.

18. Amirkhani, A., Shirzadeh, M. & Kumbasar, T. Interval Type-2 Fuzzy Cognitive Map-Based Flight Control System for Quadcopters. Int. J. Fuzzy Syst. 22, 2504–2520 (2020). https://doi.org/10.1007/s40815-020-00940-8.
19. L. Hu, J. Zhang, Y. Xiang and W. Wang, "Neural Networks-Based Aerodynamic Data Modeling: A Comprehensive Review," in IEEE Access, vol. 8, pp. 90805–90823, 2020, https://doi.org/10.1109/ACCESS.2020.2993562.
20. Kyle, Hilton, Mark Lowenberg, and Douglas Greenwell. "Comparative evaluation of unsteady aerodynamic modeling approaches." In AIAA Atmospheric Flight Mechanics Conference and Exhibit, p. 5272. 2004.
21. Kier, Thiemo M. "Comparing Different Potential Flow Methods for Unsteady Aerodynamic Modeling of a Flutter Demonstrator Aircraft." In AIAA SCITECH 2023 Forum, p. 0177. 2023.
22. Tekaslan, Hüseyin, Yusuf Demiroğlu, and Melike Nikbay. "Surrogate Unsteady Aerodynamic Modeling with Autoencoders and Long-Short Term Memory Net-works." ." In AIAA SCITECH 2022 Forum, Jan 3–7 , 2022.
23. Tai, Shang, Lixin Wang, Yanling Wang, Chen Bu, and Ting Yue. "Flight Dynamics Modeling and Aerodynamic Parameter Identification of Four-Degree-of-Freedom Virtual Flight Test." AIAA Journal (2023): 1–14.
24. Karali, Hasan, Mustafa U. Demirezen, Mahmut A. Yukselen, and Gokhan Inalhan. "Design of a deep learning based nonlinear aerodynamic surrogate model for UAVs." In AIAA Scitech 2020 forum, p. 1288. 2020.
25. Li, Kai, Jiaqing Kou, and Weiwei Zhang. "Deep learning for multifidelity aerodynamic distribution modeling from experimental and simulation data." AIAA Journal 60, no. 7 (2022): 4413–4427.
26. J. T. Rickard, J. Aisbett and G. Gibbon, "Fuzzy Subsethood for Fuzzy Sets of Type-2 and Generalized Type- n," in IEEE Transactions on Fuzzy Systems, vol. 17, no. 1, pp. 50–60, Feb. 2009, https://doi.org/10.1109/TFUZZ.2008.2006369.
27. M. F. Zarandi, R. Gamasaee and O. Castillo, "Type-1 to type-n fuzzy logic and systems," in Fuzzy Logic in Its 50th Year, C. Kahraman et al. , Eds., Switzerland: Springer, 2016, pp. 129–157, https://doi.org/10.1007/978-3-319-31093-0_6.
28. W. Pedrycz, "Fuzzy Sets of Higher Type and Higher Order in Fuzzy Modeling," in Frontiers of Higher Order Fuzzy Sets, A. Sadeghian and H. Tahayori, Eds., New York, NY, USA: Springer, 2015, pp. 31–49, https://doi.org/10.1007/978-1-4614-3442-9_3.
29. A. Mohammadzadeh, M. H. Sabzalian and W. Zhang, "An Interval Type-3 Fuzzy System and a New Online Fractional-Order Learning Algorithm: Theory and Practice," in IEEE Transactions on Fuzzy Systems, vol. 28, no. 9, pp. 1940–1950, Sept. 2020, https://doi.org/10.1109/TFUZZ.2019.2928509.
30. A. Mohammadzadeh, O. Castillo, S. S Band and A. Mosavi, "A Novel Fractional-Order Multiple-Model Type-3 Fuzzy Control for Nonlinear Systems with Unmodeled Dynamics," International Journal of Fuzzy Systems, early access ,2021, https://doi.org/10.1007/s40815-021-01058-1.
31. Z. Liu, A. Mohammadzadeh, H. Turabieh, M. Mafarja, S. S. Band and A. Mosavi, "A New Online Learned Interval Type-3 Fuzzy Control System for Solar Energy Management Systems," in IEEE Access, vol. 9, pp. 10498–10508, 2021, https://doi.org/10.1109/ACCESS.2021.3049301.
32. D. Singh, N. K. Verma, A. K. Ghosh and A. K. Malagaudanavar, "An Approach Towards the Design of Interval Type-3 T-S Fuzzy System," in IEEE Transactions on Fuzzy Systems, https://doi.org/10.1109/TFUZZ.2021.3133083.
33. F. Liu, "An efficient centroid type-reduction strategy for general type-2 fuzzy logic system," Information Sciences, vol. 178, pp. 2224–2236, 2008, https://doi.org/10.1016/j.ins.2007.11.014.
34. J. M. Mendel, F. Liu and D. Zhai, "$\alpha$ -Plane Representation for Type-2 Fuzzy Sets: Theory and Applications," in IEEE Transactions on Fuzzy Systems, vol. 17, no. 5, pp. 1189–1207, Oct. 2009, https://doi.org/10.1109/TFUZZ.2009.2024411.

35. W. Zou, C. Li and N. Zhang, "A T–S Fuzzy Model Identification Approach Based on a Modified Inter Type-2 FRCM Algorithm," in IEEE Transactions on Fuzzy Systems, vol. 26, no. 3, pp. 1104–1113, June 2018, https://doi.org/10.1109/TFUZZ.2017.2704542.
36. C. Li, J. Zhou, L. Chang, Z. Huang and Y. Zhang, "T–S Fuzzy Model Identification Based on a Novel Hyperplane-Shaped Membership Function," in IEEE Transactions on Fuzzy Systems, vol. 25, no. 5, pp. 1364–1370, Oct. 2017, https://doi.org/10.1109/TFUZZ.2016.2598850.
37. Stevens B.L and Lewis F.L. Aircraft Control and Simulation. John Wiley & Sons, 2003.
38. D. Singh, N. K. Verma, A. K. Ghosh and A. K. Malagaudanavar, "An application of interval type-2 fuzzy model based control system for generic aircraft," Applied Soft Computing, Volume 121,2022,108721, ISSN 1568-4946.
39. ATTAS aircraft flight data, Available [Online]: https://arc.aiaa.org/doi/suppl/10.2514/4.102790.
40. J. M. Mendel, Uncertain Rule-Based Fuzzy Logic Systems: Introduction and New Directions, NJ, Upper Saddle River: Prentice-Hall, 2001.
41. M. Biglarbegian, W. W. Melek and J. M. Mendel, "On the Stability of Interval Type-2 TSK Fuzzy Logic Control Systems," in IEEE Transactions on Systems, Man, and Cybernetics, Part B (Cybernetics), vol. 40, no. 3, pp. 798–818, June 2010, https://doi.org/10.1109/TSMCB.2009.2029986.

# Chapter 8
# Epilogue

This book focuses on the development of higher-order fuzzy logic systems (FLSs) and their use in solving complex real-world problems, especially in aerospace applications.

The opening chapters introduce the background of fuzzy logic, beginning with Type-1 Fuzzy Sets and moving toward Interval Type-2 (IT2) and General Type-2 (GT2) systems. The middle chapters describe the design and performance of Fuzzy Model-Based (FMB) controllers. Type-1, IT2, and GT2 FMB control systems are tested on aircraft and UAV models. Later chapters shift toward Interval Type-3 (IT3) Fuzzy Logic Systems, which are a newer advancement. IT3 systems add an extra layer of uncertainty modeling by embedding Interval Type-2 sets inside higher structures.

The book concludes by showing that fuzzy logic has grown from simple Type-1 systems to advanced Interval Type-3 models, each step adding more power to handle uncertainty. Through examples in aircraft and UAV control, it becomes clear that these methods are not only theoretical but also practical, offering stability and robustness where traditional approaches struggle. The journey from Type-1 to IT3 systems highlights how knowledge evolves, opening the way for further research and new applications in engineering and beyond.

 
D. J. Singh et al., *Fuzzy Systems with Applications to Modeling and Control of an Aircraft*, Algorithms for Intelligent Systems, https://doi.org/10.1007/978-981-95-7888-7_8

GPSR Compliance
The European Union's (EU) General Product Safety Regulation (GPSR) is a set of rules that requires consumer products to be safe and our obligations to ensure this.

If you have any concerns about our products, you can contact us on

ProductSafety@springernature.com

In case Publisher is established outside the EU, the EU authorized representative is:

Springer Nature Customer Service Center GmbH
Europaplatz 3
69115 Heidelberg, Germany

www.ingramcontent.com/pod-product-compliance
Ingram Content Group UK Ltd.
Pitfield, Milton Keynes, MK11 3LW, UK
UKHW020633260926
475829UK00003BA/59
*9789819578870*